# *Why Evolution Works (and Creationism Fails)*

# *Why Evolution Works (and Creationism Fails)*

MATT YOUNG AND PAUL K. STRODE

RUTGERS UNIVERSITY PRESS
*New Brunswick, New Jersey, and London*

Library of Congress Cataloging-in-Publication Data

Young, Matt, 1941–
Why evolution works (and creationism fails) / Matt Young, Paul K. Strode.
p. cm.
Includes bibliographical references and index.
ISBN 978-0-8135-4549-3 (hardcover : alk. paper)
ISBN 978-0-8135-4550-9 (pbk. : alk. paper)
1. Evolution (Biology) 2. Creationism I. Strode, Paul K., 1967– II. Title.
QH366.2.Y69 2009
576.8—dc22

2008040062

A British Cataloging-in-Publication record for this book is available from the British Library.



Visit our Web site: http://rutgerspress.rutgers.edu

Manufactured in the United States of America

*Paul Strode dedicates this book to his daughter, JaneMelyn, who once gave him the ultimate (however incorrect) compliment one evening during bath time after he explained why we have goose bumps: "Daddy, how do you know all things?"*

*Matt Young dedicates the book to his grandchildren, Alexandra and Noah Young, and Toby Shannon, and hopes that they do not have to write the same book a generation from now.*

# Contents

## Figures

# Tables and Boxes

Tables



Boxes

# Foreword

Some time ago after a public lecture I gave, I had the pleasure of a long and fascinating conversation with several young men who were obviously very intelligent, educated, conservative Christians. They had good questions and they listened to what responses I could offer them, and in return, they were able to ponder some of my own questions for them and provide good answers.

They asked me why, if evolution was governed by random processes, it was not completely incompatible with the idea that there was a purpose and meaning to life, one that could be directed by a divine Creator. This is a good question, but it also begs the question that evolution is in fact governed by random processes. It isn't. Natural selection, the great insight of Darwin and Wallace, is anything but random, which is also true of sexual selection, Darwin's other great evolutionary process. In fact, I don't know of any really important processes in evolution that are random (random drift is one, but we don't know how important it is; neutral evolution isn't important to evolution until something happens to it, which is usually selective). When we talk about "random mutations," we're not talking about causes or effects of mutations (anything can't happen), but about the distribution of predictable effects of mutations in populations (we can know risk factors, causes, and statistical incidences, but we can't predict in advance which individuals will be affected). For scientists, "random" is just a term that describes the statistical distribution of some known effect in a sample.

So, if evolution is not random, does it still deny a purpose or meaning to life? These really aren't questions for science, at least in the sense of ethics, aesthetics, or morals. The purpose of life is to make more life: to survive and reproduce. This is not a conscious purpose of any organisms except humans, as far as we know, but without surviving and reproducing, a species becomes extinct. Then, does life have a direction? The only direction in science is time, and we know from our studies of the evolution of life that species have evolved through time by changing and adapting to whatever the physical environment and other organisms throw at them. This history—or macrohistory, if you will—is certainly a direction, a chronicle of events and changes through time from which we derive ideas about patterns and processes. And beyond this, science does not go.

But, one of my companions asked, if you don't have the fear of losing salvation to provide the basis for your morality, what's to stop you from killing this guy next to you and taking his stuff? I really didn't know, but I replied by asking him why he would want to do that. Wouldn't he regret it? Wouldn't there be reprisals, pursuit, conviction in a court of law? Wouldn't he feel terrible for the victim's friends and family? My companions replied that they could not understand how a system of morality could be built without being based on the fear of divine retribution. So, apparently, for them the problem of good and evil is purely theological, not rational. Upon further questioning it turned out that my companions had not read Plato, and so could not conceive of the idea of the Good in a purely philosophical sense that could be put into social practice. There is also a long tradition of deriving morality from biological principles, and these are discussed in this book.

In America today, and in some other parts of the world, we face a dichotomy of approaches to the world. One is authoritarian, based on the primacy of revealed knowledge that is subject to discussion and examination (and sometimes to revision, though this is not always admitted), but not to test and refutation. This approach is religious, and is based on sacred texts whose wisdom is not questioned; if secular experience is contrary to sacred revelation, the former must be wrong, not the latter. The second approach to the world comes from the Enlightenment, and regards reason, rationality, and empirical experience as the best guides to navigating the real world—by which is meant the understanding of the natural world as well as the development of moral, social, and political precepts. It admits useful ideas from theology and religion, but not for their own sake—rather for their worth, as measured in practical terms. Knowledge is in this approach derived from experience, experiment, testing, and pragmatics. But it also assumes that the empirical world is real, is knowable, and is regulated by patterns and processes that have natural and predictable mechanisms in some sense.

We all know where these two worldviews clash; it is the subject of this book. I hope that readers will find what they need here to begin to evaluate for themselves how and to what extent each traditional approach governs and influences the decisions they make in their lives and the views that they hold. In the twenty-first century, the stakes are too high for us not to be aware of the basis of the decisions that govern our lives and the lives of others.

Kevin Padian<br>Berkeley, California<br>August 2008

# Preface

> The greater the ignorance, the greater the dogmatism.
>
> —Sir William Osler (father of modern medicine)

THIS BOOK IS BEING published on the one hundred fiftieth anniversary of the publication of Charles Darwin's *On the Origin of Species* and the two hundredth anniversary of his birth. It is an impassioned argument in favor of science—primarily the theory of evolution—and against creationism. Why impassioned? Should not scientists be dispassionate in their work? Perhaps, but it is impossible to remain neutral when, after one hundred fifty years of stunning success, one of our most successful scientific theories is under attack, for religious and other reasons, by laypeople and even some scientists who willfully distort scientific findings and use them for their own purposes.

Most educational books on evolution are written for parents or teachers and are, at best, only tangentially aimed at students. This short book, by contrast, is written primarily for college students, but it will also be useful to their parents and teachers. Nothing will make the eyes glaze over more quickly than technical jargon, and we carefully avoid such jargon wherever possible. In the same way, we stick to broad principles, hoping to make evolution clear without bogging the reader down in too much detail.

Additionally, this book addresses what other books leave out: how science works and how pseudoscience works. We demonstrate the futility of "scientific" creationism and other arguments designed to show that evolution could not have produced life in its present form. The direct descendant of scientific creationism is intelligent-design creationism, and we devote more than one chapter to debunking that misconception.

We next turn to how evolution actually works. Before the theory of evolution, natural scientists thought that they saw design in nature. We show why evolution predicts the appearance of design in nature, though not the reality of design, or at least not purposeful design. In the same vein, we discuss the problem, for creationists, of poor design and why we find it, too, in nature. Then we take up cosmological arguments offered by creationists to explain our existence and find them wanting as well. We conclude with a frank discussion of science and religion, specifically arguing that science by no means excludes religion, though it ought to cast doubt on certain religious claims that are contrary to known scientific fact.

Finally, we address this book, in part, to those who, for whatever reasons, deny what we consider well-founded scientific facts such as the antiquity of the earth and the descent of species. At the very least, they have an obligation to understand precisely what they are rejecting. We also hope to provide parents, teachers, and others with sound arguments they can easily understand and give them ammunition with which to defend modern science.

# Acknowledgments

Paul Strode thanks his wife, Sarah Zerwin, for her encouragement, suggestions, and patience, and especially for pushing him out the door on weekends: "Go! Write!" Paul is also thankful for all the great coffee that is brewed in various locations in Boulder, Colorado, especially the Bookend, Trident Cafe, and the Brewing Market, as well as for their creating an ambiance conducive to writing. When Paul read his first draft of the section on the scrotum to a Boulder High School chemistry teacher, Laura Duncan, she exclaimed, "Sounds like design by committee!" This became the title of chapter 14. Paul owes his first exposure to evolution to his freshman zoology professor at Manchester College (IN), Dr. William R. Eberly, a man of both deep faith and a deep understanding of evolution. Paul would finally like to thank his coauthor, Matt Young, for bringing him on board, for being constantly cheery, and for patiently waiting for drafts of chapters and rewrites.

Matt Young is especially indebted to Michael Grant of the Department of Ecology and Evolutionary Biology of the University of Colorado for advice and encouragement that began long before this book was a gleam in anyone's eye. He further thanks his wife and harshest critic, Deanna Young of the Colorado School of Mines, for allowing him to estivate or hibernate, depending on the season, in front of a computer as the book was in progress, and also for critiquing several chapters and suggesting thought questions. Finally, he is uncommonly grateful to Paul Strode for signing on to the project at a critical time and contributing such splendid material.

Both authors offer profuse thanks to Alyssa Wiener, a senior at Boulder High School at the time of this writing, for her excellent drawings of the Canary Island lizards, the human knee, and the human eye. We further thank Paul's 2006–2007 advanced placement biology students for reading and commenting on the first iteration of the manuscript, and his 2007–2008 advanced placement biology students for their interest and encouragement.

We are likewise indebted to G. Brent Dalrymple, Professor Emeritus of the College of Oceanic and Atmospheric Sciences at Oregon State University, who did not know us from Y-chromosomal Adam, for his careful review of chapter 15. Glenn Branch of the National Center for Science Education likewise went beyond the call of duty when he read and critiqued chapter 3 twice in succession. Mark Perakh, Professor Emeritus of Physics at California State University, Fullerton, critiqued several chapters to good effect, and Jeffrey Kieft, associate professor of biochemistry and molecular genetics, University of Colorado Denver School of Medicine, brought clarity to Part III.

Others who helped us in various ways, typically by reading one chapter or one important section, include, in no special order, Sarah Wise, then a graduate student in the Department of Ecology and Evolutionary Biology of the University of Colorado; Mark Isaak, author of the *Counter-Creationism Handbook*; Bill Jefferys, Professor Emeritus of Astronomy at the University of Texas at Austin; Thomas LaFehr, Distinguished Senior Scientist in the Department of Geophysics at the Colorado School of Mines; James A. McNeil, professor of physics at the Colorado School of Mines; Kevin R. Henke, Research Scientist at the Center for Applied Energy Research at the University of Kentucky; Tom Schneider, Research Biologist at the National Cancer Institute; John G. Brawley, amateur scientist; Chris Stassen, software engineer; Gaythia Weis, analytical chemist; Taner Edis, associate professor of physics at Truman State University; Eric Johnson, retired physicist with the National Institute of Standards and Technology; Michael Murphy of the Swinburne University of Technology, Victoria, Australia; and Ted Zerwin, University of Denver.

In addition to Mr. Branch and Professor Perakh, several contributors to the Web log "The Panda's Thumb" either read portions of the manuscript or responded to our anguished cries for help: Nicholas Matzke of the University of California at Berkeley; Jason Rosenhouse, assistant professor of mathematics at James Madison University; Burt Humburg, a hospitalist at York Hospital, York, Pennsylvania; Tara C. Smith, assistant professor of epidemiology at the University of Iowa; P. Z. Myers, associate professor of biology at the University of Minnesota, Morris; Ian Musgrave, a molecular pharmacologist and senior lecturer at the University of Adelaide, Australia; and Larry Moran, professor of biochemistry at the University of Toronto.

Any mistakes in the book are entirely our own, and we bear full responsibility for them. Deep down, however, each author suspects the other.

Finally, we are indebted to our editor, Doreen Valentine, for promptly accepting the proposal and graciously working with us for approximately two years as the proposal grew into a book. Two anonymous reviewers engaged by Rutgers University Press provided excellent suggestions for improvement. Rick Delaney not only very capably copyedited the manuscript, but also made many insightful suggestions regarding the content and clarity of the book. Marilyn Campbell and Suzanne Kellum shepherded the book through production and responded to our quibbles promptly and cheerfully.

PART ONE

# *The Basics and the History*

CHAPTER 1

# Introduction

> Evolution is so clearly a fact that you need to be committed to something like a belief in the supernatural if you are at all in disagreement with evolution. It is a fact and we don't need to prove it anymore. Nonetheless we must explain why it happened and how it happens.
>
> —Ernst Mayr, Harvard University

THE THEORY OF EVOLUTION is one of the most successful in all of science. Its predictions have been verified countless times since the publication of Charles Darwin's *On the Origin of Species* in 1859. As we will see, the evidence is so stunning and complete that we are confident in saying that *descent with modification* is an observed fact. Further, the modern theory of evolution, also called the *modern synthesis*, combines Darwin's concept of descent with modification with the theory of genetics. The modern synthesis accounts for the observed facts better than any other theory.

Biological, or organic, evolution refers to descent with modification, or changes among organisms over generations. The unit of evolutionary change is the population, not the individual. But change begins with the molecules and molecular interactions of genetic material in an individual and ultimately develops in individuals physical or behavioral traits that differ from their ancestors'. Because the change is in the genetic material of the individual, that change is heritable. In other words, if the individual reproduces successfully, that change is passed on to its offspring. Ultimately, the change may become common within the population of which the original individual was a member. At this point, we conclude that the population has evolved, because it has experienced a widespread biological change. Indeed, the definition of evolution is the widespread change of a genetic trait within a population. Finally, if the population evolves so far that it cannot normally interbreed with the ancestral population, we say that a new species has evolved.

## Why Is Evolution Important?

Evolution is critically important today, yet many people—especially in the United States—reject it, often on religious grounds. Why is it so important, and why do so many people reject it? We will answer those questions in order.

The world that the next generation inherits will be shaped in large part by our understanding of evolution or, perhaps, by our lack of understanding. For example, when antibiotics were discovered, many physicians wrongly assumed that bacterial diseases would become a thing of the past. Only a few prescient thinkers recognized that bacteria would evolve resistance to antibiotics and warned against overusing them. Such warnings went unheeded even though a penicillin-resistant strain of the bacterium *Staphylococcus aureus*, which can cause skin infections, meningitis (a brain disease), pneumonia, and septicemia (blood poisoning), was discovered within four years of the first mass production of the very first antibiotic. Indeed, today, a number of harmful bacterial species, including those that cause tuberculosis and childhood ear infections, are resistant to at least one widely prescribed antibiotic. These organisms have evolved into their new, antibiotic-resistant forms by random genetic mutation, the process that provides the raw material for natural selection. *Plasmodium*, the parasite that causes malaria, has likewise grown resistant to the early treatment, quinine, and its later derivatives.

Because there are only a limited number of ways to kill bacteria (without also killing the host), we may have few remaining opportunities to develop new, effective antibiotics. If public health policy reflected a keen understanding of evolution, we would stop prescribing antibiotics for diseases against which they are ineffectual, and we would not administer them routinely to livestock and fowl, as we do now.

Similarly, many insects and noxious weeds have become resistant to insecticides and herbicides, and farmers have to use more and stronger chemicals to control them. In addition, the mosquitoes that carry malaria have become resistant to insecticides. Besides mosquitoes, many species of lice, ticks, bedbugs, and fleas, not to mention cockroaches, have become resistant to one or more insecticides. Many of these insects transmit diseases to humans. Any evolutionary biologist could have predicted insecticide resistance.

### *Erosion of Genetic Diversity*

An evolutionary biologist would also be greatly concerned about the loss of diversity among our food crops and the lack of diversity in our agricultural fields. Today, for example, farmers in developed countries plant only a relatively few varieties of wheat. They used to plant hundreds of varieties of apple; now we commonly see less than a dozen. A potent fungus attacking one of our more common food crops could wipe out a significant fraction of our food supply.

Indeed, 30 percent of all farm animal breeds worldwide may be at risk of extinction. Ironically, these animals have evolved special traits that have been naturally selected to make them well adapted to local environmental challenges like disease and drought. Nevertheless, these local breeds are being replaced by farm animals that originated in developed countries with temperate climates and have been bred for high milk production and fast growth. Livestock from the developed countries may be genetically unfit to survive and reproduce in areas like sub-Saharan Africa, where they may not be well adapted to the local climates or grasses. As the new breeds with their low diversity and higher yields become more prevalent, the local populations decrease and are put at risk of extinction, along with the special traits and the genes that control them. This is an example of an evolutionary phenomenon called the *erosion of genetic diversity*. When genetic diversity is limited, plants and animals may not be able to respond to a new epidemic because the genes that might have helped them respond are gone.

We also see erosion of genetic diversity in wild populations of plants and animals, but for different reasons. Human activity has caused the degradation and in a few cases the complete disappearance of many (perhaps most) of the Earth's ecosystems. As natural habitats are altered and removed for human benefit, populations of species that depend on these habitats shrink, and some populations become so small that they can no longer survive. Conservation biology, an entirely new branch of biological science, was developed as a reaction to this biodiversity crisis. One of the ethical and philosophical foundations of the Society for Conservation Biology is that evolution is essential to the continued existence of life on Earth; interference with evolutionary patterns disrupts interactions among species and therefore reduces general ecosystem health. Human interference could result in such severe losses of biodiversity that the very nature of our existence will be put at risk. Only through a deep understanding of the mechanisms that drive evolution can we expect to be effective stewards of the organisms upon which we depend.

A theme that connects the examples in the preceding paragraphs is economics. Indeed, the thirst for economic growth influences the decisions we make about our effects on the natural world, often regardless of our knowledge and understanding of evolutionary processes.

### *Evolution in Medicine*

If a virus or a bacterium can be transmitted easily from one human to another, it is, so to speak, in the interest of that virus to not harm its hosts. The common cold, for example, is transmitted when people are in close if not direct contact. The virus does not live long in the air, so person-to-person contact is necessary for the health of the virus itself. The virus thus thrives only if an infected person can get up and go to school or work. A mutant virus that forces a person to stay at home is at a disadvantage with respect to a virus that does not, so the common cold has evolved to *not* make people very sick.

Similarly, a great many people, including almost all adults over thirty years old, are infected with the Epstein-Barr virus (EBV). This virus quietly takes over the machinery that controls cell division, yet produces almost no symptoms. Such viruses can lie dormant for years without being detected. Only when college or high-school students who have already been infected with EBV, for example, get run down does their immune system lose control, and they get mononucleosis. Worse, when something goes terribly wrong, the presence of the *pathogen* (an organism that causes disease) becomes a serious problem for its host. For example, malaria patients have a compromised immune system, and their bodies lose their ability to control cell division in the cells infected with EBV. Cells divide rapidly, and soon a tumor forms. The patient then can develop a form of cancer called Burkitt's lymphoma. Burkitt's lymphoma is common in equatorial Africa, where malaria is also common.

If a pathogen does not have to be transmitted by direct contact, then it can kill its host quickly and still survive by infecting another host. Malaria, for example, is spread from person to person by mosquitoes. That is, a mosquito bites an infected person and swallows some parasites. That mosquito bites another person and inoculates that person with the parasites. The disease is spread without direct contact between people. It is therefore in the interest of the parasite to strike its victims hard and fast: People infected with malaria are less able to defend themselves against mosquitoes. They may be bitten many times, so the parasite spreads effectively. Evolutionary biology tells us that, if we can keep critically ill people from being bitten by mosquitoes, as with mosquito netting, then mosquitoes will be able to bite only relatively healthy people. Hence, as with the common cold, the virulence of malaria will decrease.

A waterborne disease such as cholera is also spread without direct contact. Cholera causes diarrhea. If sanitation is poor, then sick and dying people will transmit cholera by contaminating the water supply. We can attack the disease by cleaning up the water supply and by ensuring that water for washing clothing and bedding is not mixed with the drinking water. When we do so, we find that the bacterium that causes cholera becomes less virulent, because now it must keep its host alive if it is to survive. People still get cholera, but the disease is not as serious as it was in the eighteenth and nineteenth centuries. Pathogens adapt to new conditions, and the study of evolution gives us clues on how to fight certain diseases.

Genetic diseases may also be understood by applying evolutionary theory. Sickle-cell anemia, for example, is a genetic disease that is most prominent among people of sub-Saharan African descent. Sickle-cell anemia evolved as an accidental replacement of one molecule for another in the DNA within the gene family that makes hemoglobin proteins. As a result, one of the main proteins in hemoglobin becomes horribly misshapen, compromising the oxygen-carrying capacity of hemoglobin. The misshapen protein can also easily form crystals within red blood cells, causing them to become elongated, or sickle-shaped. These elongated cells not only have a reduced oxygen-carrying capacity; they also block blood vessels and cause pain or organ failure. Before the development of modern medicine, patients with sickle-cell anemia had considerably shortened lives.

Individuals who carry only one copy of the sickle-cell gene do not express the symptoms of the disease. Rather, those who express the symptoms of the disease carry two copies of the gene. In part because of their shortened lives, they are less likely to produce offspring than those who do not express the symptoms. We might therefore think that a disease-causing mutation as debilitating as sickle-cell anemia would quickly disappear from the human population. Why then has this disease persisted? Precisely because an individual needs two copies of the sickle-cell gene to have full-blown sickle-cell anemia. What is truly fascinating from an evolutionary perspective is that *carriers*, or individuals who carry one normal copy of the hemoglobin gene and one sickle-cell copy, are almost completely protected from the symptoms of malaria. In the presence of malaria-carrying mosquitoes, carriers are more fit than individuals who carry only normal hemoglobin genes; to the population as a whole, the advantage of being resistant to malaria outweighs the disadvantage of sickle-cell anemia, so the sickle-cell gene persists. Without an understanding of evolution by natural selection, we would have no way of explaining why sickle-cell anemia does not disappear in a few generations.

*Genealogy*

Finally, we note the popularity of genealogy; many people study genealogy as a hobby and take it seriously because it puts their lives into a larger historical context. People want to draw a genealogical chart and work it back through as many generations as they can. On the other hand, some geneticists and evolutionary biologists approach genealogy not as a hobby, but as a scientific pursuit. They ask interesting questions and often come up with fascinating answers. For example, in 2003, Tatiana Zerjal of the Department of Biochemistry at the University of Oxford, along with a group of other scientists, published in the *American Journal of Human Genetics* a paper that explains genetically the genealogy of Genghis Khan and his descendants. They show that the Y-chromosomes of 8 percent of the males throughout a large part of Asia most likely originated with Genghis Khan himself. Indeed, this finding means that 0.5 percent of males (around 16 million) in the world may be descended from Genghis Khan. Historians confirm this possibility with their accounts of the conquest and slaughtering of many populations as

Genghis Khan and his male descendants established the largest land empire in history. At the same time, Genghis Khan and his descendants fathered hundreds of children.

Evolutionary biology allows us to extend our genealogy much farther back. For example, analysis of the DNA of females leads back to a single female, nicknamed "mitochondrial Eve." Importantly, mitochondrial Eve is not the first woman who ever lived, but rather the individual woman whose lineage has survived to the present. The descendants of most, if not all, other females alive at the time of Eve appear to have died out. If that is so, then we are all descended from Eve. Similar analysis of the DNA of males leads to a single male, nicknamed "Y-chromosomal Adam," who is probably the ancestor of all males alive today. Oddly, Adam and Eve did not live at the same time. (Eve most likely lived 140,000 years ago and Adam, 60,000 years ago; it is safe to say that they never met.) Moreover, we can project back approximately 4 billion years to the Last Universal Common Ancestor, or LUCA, from which descended the three major domains of life, Archaea, Bacteria, and Eukarya.

### WHY DO PEOPLE DENY EVOLUTION?

We sometimes hear people say that scientists "believe in" evolution, as if evolution were a religious *dogma*, that is, a doctrine declared to be authoritative by a church or other religious organization. Scientists do not believe in evolution in the way that some theologians believe in dogmas. To the contrary, no matter what their religion, scientists *accept* evolution on the weight of the evidence. Specifically, most scientists (and the vast majority of biologists) find the physical evidence in favor of descent with modification to be conclusive and the modern theory of evolution to best explain that fact. Acceptance of evolution does not depend on any particular religion or nationality; biologists the world over are virtually unanimous in accepting and working with evolution.

If biologists are certain of the validity of evolutionary theory, why do so many lay persons, especially in the United States, reject it? Probably for several reasons: They confuse evolution with atheism or think it may lead to atheism; they think it contradicts the Bible; or they think it implies a purposelessness to their existence. Others think that evolution cannot explain morality and indeed leads to immorality. Additionally, many people simply do not understand the theory and incorrectly think that chance is its sole explanation for the evolution of complex organisms.

#### *Evolution and Unbelief*

The charge that evolution may lead to atheism is exaggerated but not wholly unfounded. Several surveys taken between 1914 and 1998 have shown, for example, that biologists and anthropologists are more likely to disbelieve in

God than are physical scientists or engineers. No one is certain why, but possibly it is because biologists and anthropologists see a great deal of evil (or the animal kingdom's equivalent of evil) in their studies of animal behavior, and they think that evil militates against the existence of God. For example, many species of insects employ a strategy called *parasitoidism*. One such group includes species in the wasp family Ichneumonidae. Ichneumon wasps lay their eggs in the bodies of caterpillars so that their larvae can eat the caterpillars alive. Darwin himself was so disturbed by the ichneumon wasp that he wrote in a letter, "I cannot persuade myself that a beneficent and omnipotent God would have designedly created the Ichneumonidae with the express intention of their feeding within the living bodies of Caterpillars, or that a cat should play with mice." Importantly, Darwin did not say "God," but "a beneficent and omnipotent God."

Such considerations, we argue, do not necessarily rule out a deity, though they may push the concept of a personal, intervening deity to arm's length.

But suppose that evolution did in fact imply atheism? Would that necessarily bear on the truth or falsity of evolution? No. To understand that statement, we discuss a logical fallacy that philosophers call the *appeal to the consequence*: rejecting a claim of fact because its consequences are perceived to be undesirable. Thus, it is a fallacy to reject evolution because you do not like its consequences (or what you think are its consequences). Evolution may or may not lead some people to disbelieve in God; that has nothing to do with its validity. In the same way, a false doctrine that leads people to believe in God is no more true for that consequence. In short, a claim of fact must be judged solely on its merits, not on our preference for its consequences. More to the point, evolution must be judged on its merits, even though it may lead some people to disbelieve in God.

A related objection to evolution is its apparent discord with the Bible, especially the first chapter of Genesis. Although some people have tried to reconcile the account in Genesis with the modern theory of evolution, those arguments are inconsistent and unconvincing: The theory of evolution disagrees with much that is written in the Bible—but only if we interpret the Bible literally. A great many religious people take the Bible figuratively or as spiritual metaphor (or distinguish between those parts of the Bible that relate history and those that do not) and find no contradiction between their religion and the theory of evolution. We do not presume to offer religious advice, but we urge *biblical literalists* to read further and be open to a thoughtful consideration of evolution.

*Evolution and Purpose*

Some people deny evolution because they think that the universe must have some overall purpose, or else life is futile. Does evolution imply purposelessness? Not to the many religious believers who accept it. Such adherents range from *deists*, who think that God set the universe in motion and then wholly or largely withdrew, to *theistic evolutionists*, who think that God is active in the world today but does not intervene to contradict *natural law*. To most theistic evolutionists, evolution is God's way of fulfilling God's purposes, whatever those may be. This book has no quarrel with religious believers such as theistic evolutionists; rather, its quarrel is with creationists who flatly deny the findings of modern science.

But, as we noted in connection with atheism, even if evolution implied a lack of purpose, it would be a logical fallacy to reject it on that grounds. Evolution is a fact, whether or not we like the consequences. If that fact implies a lack of purpose—and not everyone concedes that it does so—then people will simply have to find their own purpose or their own meaning in life.

*Evolution and Morality*

Yet another objection to evolution is that it is amoral or leads to immorality. Besides being another appeal to the consequence, the claim is not true. To the contrary, as we show in chapter 17, there is good evidence that morality is an evolved behavioral characteristic and that it derived from the propensity of organisms—from slime molds to ants to pack animals such as wolves—to cooperate among their own species. Religious leaders have often exemplified good morality or even been enforcers of public morality, but we do not need religion to be moral; it is arguably in our nature.

Similarly, evolution has been blamed for *social Darwinism*, the theory that those who have wealth and power do so because they are biologically superior, and its odious cousin, *eugenics*, which proposes that only certain supposedly superior groups of people are fit to produce offspring. Social Darwinism circularly argues that if some people are successful, then they must be superior. It has been used to justify, for example, the laissez-faire social and economic policies of the robber barons of the nineteenth century. Social Darwinism confuses what is with what should be. Exploitation, slavery, and the divine right of kings have existed for thousands of years and did not need to rely on evolution for theoretical support. Blaming evolution for social Darwinism is yet another appeal to the consequence.

*Chance in Evolution*

Many find the theory of evolution distasteful because they think it relies entirely on chance. They liken the development of a multicellular organism to tossing a million coins and getting just the right outcomes or tossing all the parts of an airplane into the air and having them fall to earth as a completed Boeing 747. No one seriously thinks that evolution works that way, and, indeed, the Boeing 747 evolved to its present form beginning with the Wright brothers' plane at Kitty Hawk through a great many intermediate forms. We will show later in the book that, although chance plays a role in evolution, natural selection is by no means wholly random.

Consider the example of the shark and the dolphin. Both live in similar environments and are adapted for swimming underwater. Both swim by waving or oscillating their tails, the shark horizontally and the dolphin vertically. Even though the shark gets its oxygen from water and the dolphin gets it from air, they have remarkably similar shapes; we would say they are both well *adapted* to the liquid environment. Yet the shark evolved from a fishlike ancestor, and the dolphin evolved from a land mammal. Chance played a role, for example, in how fast and by what route each evolved into its present form, but the needs imposed by the watery environment dictated the streamlined shape of each animal. Evolution thus displays a sort of lawlike behavior and does not depend wholly on chance. The history of each lineage, for example, determined how each animal solved certain problems, such as locomotion.

Natural selection often starts with different raw materials and converges to the same solution. The constraints imposed by the environment forced the shark and the dolphin to develop similarly streamlined shapes, even though they are not closely related. These are not isolated examples but exemplify a common theme, *convergent evolution*, the evolution of similar traits in response to similar environmental constraints. Still, their different histories forced the dolphin to swim with an up-and-down undulatory motion, whereas fish swim with a side-to-side motion. Evolution does what it can with the raw materials available to it; because the terrestrial ancestors of the dolphin could not bend their backs in a side-to-side motion, the dolphin does not swim like a fish.

The Tasmanian wolf, or thylacine, is a marsupial native to Australia and New Guinea. It has recently become extinct and is preserved only in paintings and photographs. It is more closely related to a kangaroo and a possum than to a wolf, but it looks remarkably like a wolf and fills the same ecological niche as the wolf. And other extinct marsupial, *Thylacosmilus*, closely resembled the saber-toothed tiger but is only very distantly related. The echidna, or spiny anteater, is closely related to the platypus, but it has developed a long snout and a long, sticky tongue like those of the true anteaters (which are related to sloths) and aardvarks. In addition, the echidna has evolved spines or quills similar to those of hedgehogs and porcupines, none of which are closely related to each other. Such convergent evolution is not the result of chance.

*Camouflage* and *mimicry* differ from convergent evolution in that the organism evolves to look like the background in which it lives (camouflage) or to physically resemble some other organism (mimicry). Like convergent evolution, camouflage and mimicry do not evolve by chance but gradually, owing to the needs of the organism in its environment. Thus, to avoid predators, the snowshoe hare develops a white coat in the winter, the praying mantis looks like a leaf, and the walking stick (an insect) looks like a twig. Similarly, predators such as the polar bear and the Arctic fox have evolved white coats to avoid detection by their prey. More interestingly, a moth may evolve to resemble a wasp, so that predators that have been stung by a wasp will avoid the moth as well as wasps, or an edible butterfly may mimic an inedible but unrelated species. These last two examples are of a special kind of mimicry called *Batesian mimicry*, after the English naturalist Henry Walter Bates, in which an innocuous species (the mimic) resembles something poisonous or dangerous (the model) in order to protect itself.

Convergent evolution, mimicry, and camouflage do not evolve by chance; they evolve gradually and for specific reasons over many generations as organisms become more streamlined, better camouflaged, or better mimics as the result of natural selection.

The well-known case of the peppered moth in England provides iconic evidence of the evolution of camouflage as environmental conditions changed. After the beginning of the Industrial Revolution, trees in and around many English cities became covered with soot. Against the black background of the soot, the light-colored moths were easily snapped up and eaten by birds. A darker variety of the same moth almost completely replaced the light-colored, peppered variety. When clean air acts were introduced and the soot gradually disappeared from the trees, the lighter variety of moth reasserted itself. Chance played a role in the selection of the darker moths, but their selection was virtually inevitable as the trees darkened and provided less and less camouflage to lighter moths. Recently, in response to creationist criticisms of the pioneering field work on the peppered moth by the naturalist Bernard Kettlewell in the 1950s, Michael Majerus of the University of Cambridge performed a multiyear experiment designed to answer the creationists' criticisms and found that the original experiments hold up very well.

*Age of the Earth*

Darwin developed his theory at a time when the age of the earth was in dispute, and William Thomson, later Lord Kelvin, had incorrectly calculated the earth to be on the order of 100 million years old at most. Darwin's theory of evolution predicted a much older earth, as did certain geological observations. As it turned out, Darwin was correct and Kelvin wrong: Kelvin had assumed that the earth was formed as a ball of molten rock and estimated its age by calculating its cooling rate. Kelvin's method was legitimate, but the calculation is almost impossible. His incorrect assumption that core of the earth was solid, however, invalidated his conclusion.

Darwin and the geologists were thus vindicated in their belief that the earth was much older than 100 million years. Radioactive dating and other techniques have since established the age of the earth as 4.5 billion years.

Many but not all of those who attack evolution believe in a young earth and a young universe, typically 10,000 to 20,000 years old but sometimes as little as 6,000 years old. Evidence from geology, astrophysics, physics, and cosmology, as well as biology, confirm conclusively that the earth and the universe are billions of years old. Thus, *young-earth creationism* becomes an attack on all of science; the attack on evolution is only the thin edge of the wedge.

*Old-earth creationists* and *intelligent-design creationists* accept the antiquity of the earth and the universe but argue, for different reasons, that biological evolution could not have taken place without at least some guidance. We will examine their arguments in detail later, but we note here that, if we accept the claims of creationists, we will ultimately stop doing science, because "at least some guidance" assumes in advance that we cannot answer many scientific questions about evolution, human origins, or indeed the origin of life. That is not to say that science and religion cannot coexist. Creationism, however, by assuming the answers, rules out investigations of certain open problems. Creationists' claims are therefore subtle attacks on biological evolution in particular and science in general. All creationists, in varying degrees, deny both the conclusions of science and the methodologies of science.

### Thought Questions

1. As we systematically erode the genetic diversity of our food crops by selecting specific, desirable traits, we also select undesirable traits in our weeds. From an evolutionary perspective, why do you think that weeds, in particular, can grow back after being pulled out by the roots or sprayed with herbicide? Why can weeds grow in dry, barren soil where food crops cannot? Why are our food crops susceptible to insect damage, whereas weeds and flowers are less so? Hint: Why do you think many plants are poisonous? Explain your answers.
2. Think of something you believe strongly. List the reasons you believe it. Are they legitimate reasons? What makes them legitimate reasons?
3. Think of something you *dis*believe strongly. List the reasons you disbelieve it. Are they legitimate reasons? What makes them legitimate reasons?
4. Citing an authority to support a contention is called an *appeal to authority*. When is it appropriate to accept the word of an authority and when not? Who can fairly be called an authority? Can you give examples of people who may be authorities in one subject but not in another?

CHAPTER 2

# *The Structure of This Book*

> One cannot help but be in awe when contemplating the mysteries of eternity, of life, of the marvelous structure of reality. It is enough if one tries merely to comprehend a little of the mystery every day.
>
> —Albert Einstein

A MODEST FRACTION of this book is devoted to refuting all varieties of creationism. Hence, we discuss, in chapter 3, a "History of Creationism and Evolutionary Science in the United States." Before the development of modern science, nearly everyone was a creationist, in a sense. But even those who believed in the Bible literally saw the necessity to interpret it so that their understanding of the Bible was consistent with scientific facts as they were understood at the time. Thus, when the antiquity of the earth became apparent, theologians proposed, for example, that the days of the creation might have been eons and not literal twenty-four-hour days. When the theory of evolution posed a new threat, it became one factor that brought about a more literal interpretation of the Bible. In the mid-twentieth century, the little-known pseudoscience of flood geology evolved into creation science, and its proponents engaged in political activity designed to force creation science into the public schools alongside evolutionary biology. When courts deemed creation science to be religion, not science, it evolved again, this time into intelligent-design creationism.

In chapter 4, we show "How Science Works." In a nutshell, scientists make hypotheses and then compute or deduce specific, detailed consequences of those hypotheses. They then carry out experiments or make observations to test whether the consequences of their hypotheses are correct. If so, then the hypotheses are tentatively accepted. More experiments or observations and more hypotheses follow in order to verify or refine the original hypotheses. Sometimes an observation or experiment precedes a hypothesis. Some hypotheses are rejected, others accepted. When enough hypotheses have been accepted and we have a large body of knowledge, then we say that we have a successful *theory*. Additionally, by performing experiments, making observations, and testing hypotheses, science constantly uncovers new facts and is fruitful in a way that pseudoscience and creationism are not.

In science, a theory is a comprehensive body of knowledge, hypotheses, and deductions that explain a broad range of facts. It is often mistaken to be a mere notion, an untested idea, as when someone says, "I have a theory about why they lost the game." In general, a mature theory is so complex and has so many interwoven strands of evidence that we cannot ask whether a theory is correct or incorrect. Rather, we may ask whether a given hypothesis can be tested. Sometimes, a hypothesis is rejected, but that does not mean that the overall theory is wrong, only that some aspect needs further investigation. Most important, we do not reject a well-supported theory just because we have found some flaws. And—creationists must understand this point—we do not accept a rival theory by default but rather subject it to scrutiny by proposing specific hypotheses, deducing consequences, and testing those hypotheses. To date, no competitor to the theory of evolution has been subjected to such scrutiny.

It is very hard to define what is science and what is not; the boundary between science and nonscience is often fuzzy. It is perhaps easier to distinguish between science and pseudoscience, as we do in chapter 5, "How Pseudoscience Works." We note specific properties that make an endeavor pseudoscience: untestable or far-fetched hypotheses; denial of known fact; intention to show that something is true, rather than find out whether it is true; and sometimes the belief that the establishment is conspiring against the pseudoscientist. In chapter 5, we expand on these themes with reference to two specific pseudosciences: homeopathic medicine and astrology. As we will see, pseudoscientists "know" in advance the answer they are trying to get and will not be swayed by facts or logic.

In chapter 6, "Why Creationism Fails," we look at classical scientific creationism, that is, the predecessor to intelligent-design creationism. Young-earth creationists believe that the universe is approximately 10,000–20,000 years old (though some still stick to the biblical age of 6,000 years) and that Noah's flood was real and worldwide. They account for the stratification we find in the fossil record by citing unspecified hydrodynamic forces and also by appealing to an unsupported theory that more-advanced animals survived the flood longer than less-advanced animals and so were deposited on top. They provide no evidence, however, that their theory makes better predictions than standard geology. Old-earth creationists accept the antiquity of the earth and the universe but try to force the chronology of the Book of Genesis to match the geological and paleontological records. They claim, for example, that human beings predated Adam and Eve, but Adam and Eve were the first humans to have a soul. Their claim is untestable because the fossil record does not distinguish between people with souls and people without.

Chapter 7, "The Argument from Design," surveys a venerable but now discredited philosophical argument that goes back at least to the Roman philosopher Cicero. The modern form of the argument from design may be traced to William Paley, the naturalist and theologian who developed the famous watchmaker analogy. Paley's argument was, in essence, that we can recognize design in an artificial mechanism such as a watch; how then can we not recognize design in the intricate workings of nature? Paley's argument was accepted as sound until Darwin and Alfred Russel Wallace developed the theory of descent with modification, which showed how the appearance of design could accumulate over eons of trial and error. Indeed, Darwin himself accepted Paley's argument until he and Wallace discovered a better explanation.

Creationism is now packaged as *intelligent design*, but it is the same old creationism dressed up in a cleaned and pressed nineteenth-century lab coat. Chapter 8, "Why Intelligent-Design Creationism Fails," outlines the modern arguments. The mathematician William Dembski considers that a very complex entity such as a chlorophyll molecule could not have arisen by chance but must have been designed. To distinguish between design and chance, he proposes an *explanatory filter*, but does not give enough detail to make the filter useful. In addition, the filter suffers from *false positives*; that is, it can sometimes detect design incorrectly. We discuss the explanatory filter in detail in chapter 8.

Like Paley, modern creationists rely on analogies between human artifacts and, for example, biological systems. The molecular biologist Michael Behe likens the bacterial flagellum to an outboard motor. He argues that since it has three interacting parts, every one of which is crucial to its operation, and since no two of the parts could function without the third, the flagellum could not be the product of evolution but rather must have been designed. It is simply too unlikely, says Behe, that all three would have assembled at one time. Biologists recognize, however, that evolution sometimes gradually adapts for one function parts that were originally used for another; we have a fair idea how the bacterial flagellum could have evolved from a flagellum that was originally used for parasitism or secretion rather than swimming. Indeed, an earlier biologist, the Nobelist Hermann Muller, in 1918 explicitly predicted the evolution of complex, interlocking systems in which parts that were once merely important evolved to become crucial. That is, Muller used evolutionary theory to predict a phenomenon that Behe eighty years later claimed could not evolve by natural selection.

Chapter 9, "The Father of Evolution," begins Part II, "The Science of Evolution," with a brief romp through the personal evolution of Charles Darwin from a promising clergyman to a paradigm-shifting force in the natural sciences. We continue in chapter 10, "How Evolution Works," with an explanation of some of the ways evolution works. First, we note that some organisms are more *fit* than others; that is, they are better able to survive and reproduce. The fitness of an animal can depend on many factors, such as weight and size, and there may be an optimum weight and size for any given species. If, however, the environment changes, the optimum weight and size may change too. The animals immediately become less fit than they were before, but their weight, size, and other characteristics may change gradually, over many generations. If the species changes enough, it may come to be regarded as a new species. Indeed, in chapter 10, we outline field observations suggesting that a certain kind of fly has evolved into a new species owing to the introduction of the apple into North America.

Chapter 11, "Recapitulation," explains how eighteenth-century embryologist Ernst Haeckel's enthusiasm for Darwin's theory resulted in some misleading drawings and decades of textbooks with the faulty idea that "ontogeny recapitulates phylogeny." Haeckel's idea of recapitulation does, however, explain some benchmarks that are essential for proper continued development of vertebrate embryos.

Next we turn to a new field of biology called evolutionary developmental biology (*Evo Devo* to its practitioners). In this exciting field, scientists investigate how genes control development and how tweaking the genes and their protein products has resulted in impressive and surprising changes in the final form and function of organisms. In chapter 12, "Evo Devo: How Evolution Constantly Remodels," we introduce the fruit fly, one of the model organisms scientists have relied on for understanding everything from the evolution of embryonic development to the human disease of cancer. We also explain how evolutionary pathways are predictable when different populations experience the same environmental pressures. We present a twist to the idea of the "survival of the fittest," and instead offer the new idea, the "arrival of the fittest." We end the chapter with a detailed explanation of how complex systems evolve, and provide evidence against the failed idea of irreducible complexity.

Chapter 13, "Phylogenetics," takes the reader through the history of the ever-changing branching models of life. Here we explain why these tree-like models are best guesses, or hypotheses, that incorporate the latest available data. Through several illustrations and explanations we show how, with new data and new ways of looking at the relationships among species, we get closer and closer to how groups of past and present animals are related. We then examine the process of building a phylogenetic tree from scratch using data from a population of lizards on the Canary Islands.

Chapter 14, the last chapter in the "Science of Evolution" unit, is titled, "Design by Committee: The Twists, Turns, and Flips of Human Anatomy." Here we investigate several aspects of being human that present our species with unique challenges. We explain that the evolutionary pathways humans have taken have resulted in some innovations that are far from perfect: evolutionary engineering is costly and also cumbersome.

Following our discussion of evolutionary biology, we begin Part IV, "The Universe." The ancient Greeks realized that fossils were the remains of plants and animals that had lived long ago; the European idea of a young earth is simply a result of a too-literal interpretation of the Bible. By the time Darwin wrote *On the Origin of Species*, geologists had begun to amass considerable evidence favoring an ancient earth. Some physicists, however, thought that the earth was much younger. As we show in chapter 15, "How We Know the Age of the Earth," the discovery of radioactivity in 1896 invalidated the physicists' arguments and also provided a stable clock with which to measure the time since the earth formed. Early attempts at *radiometric dating* were imprecise, in part because no one knew the concentration of radioactive material in a mineral when it formed. By the second half of the twentieth century, however, geophysicists had developed methods to account for that unknown and have measured the age of the earth very precisely. We can say almost without doubt that the earth's crust solidified at least 4.4 billion years ago. Similarly, cosmologists have observed the expansion of the universe and discovered the *cosmic background radiation*, which we may think of as the afterglow of the *big bang*. Using detailed theories to explain the observed properties of the background radiation, cosmologists have deduced that the universe as a whole is approximately 13.7 billion years old, with an uncertainty of approximately 1.5 percent.

As far as we can tell, the universe is not especially hospitable to life, or at least not to life as we know it. Even the solar system as a whole does not seem hospitable; the earth is most probably the only astronomical body in the solar system where complex, multicellular life has developed. We know next to nothing about the rest of the universe, except for the relatively recent discovery that a significant fraction of stars may have planetary systems. It thus seems likely that the conditions necessary for the development of intelligent life have arisen on earth purely by chance: if it had not been earth, then it might have been somewhere else.

Nevertheless, some physicists and astronomers persist in arguing that the universe was designed for life, and they point to the supposed *fine-tuning* of the fundamental physical constants as evidence. Specifically, as we discuss in chapter 16, "Is the Universe Fine-tuned for Life?," they claim that if any one of a dozen or two *fundamental constants*, such as the charge on the electron and the mass of the proton, were changed, then the universe would not be able to support complex life. The fine-tuning argument presumes life like ours, but there is no justification for such a presumption. In addition, it has been tested by a mathematical simulation, which suggests that the fundamental constants are not as finely tuned as is claimed.

The *anthropic principle* builds on the fine-tuning argument and argues that the universe was designed for us, and that we know that because we are here.

We find that argument circular and think that it is also based on a poor understanding of probability. A more recent argument infers design from the claim that life is rare but that the earth is located in a privileged position within both the galaxy and the solar system. We find this argument not only circular but also untestable because its authors will also infer a designer if life is found to be plentiful.

Part V concerns "Evolution, Ethics, and Religion." Francis Collins, the head of the Human Genome Project, is not alone in thinking that morality is uniquely human and requires a supernatural explanation. We show in chapter 17, "Evolution and Ethics," however, that cooperation has evolved at all levels in the animal kingdom and can lead to *reciprocal altruism* of the "you scratch my back, I'll scratch yours" variety. Human morality may have evolved from mere reciprocal altruism to true, selfless altruism as the result of both biological and cultural factors. The fact that ethics or morality is an evolved trait, however, by no means implies that evolutionary biology is incompatible with religion.

In chapter 18, we show "Why Science and Religion Are Compatible." Specifically, we argue that science does not exclude religion, though it surely calls certain specific religious beliefs into question. Religious beliefs that do not deny known scientific fact, by contrast, are not incompatible with science, and many scientists are themselves religious. We analyze the example of two brothers-in-law, one a palaeontologist and one a minister, each of whom began his career as a biblical literalist. Each modified his beliefs—the palaeontologist re-evaluated the evidence for evolution, while the minister carefully re-examined the internal inconsistency of the biblical accounts. We conclude the chapter by refuting the well-known contention of the palaeontologist Stephen Jay Gould that science and religion have authority over separate domains of knowledge, or *magisteria*. We argue that Gould may be right in principle, but wrong in practice, and that there must invariably be some tension between science and religion.

We conclude with the observation that evolutionary biology explains the observed facts better than any competing theory; indeed there is no competing theory. Creationism is not just a trivial pseudoscience but attacks the very foundation of modern science. It has no program but rather postulates that it can overthrow evolutionary biology, cosmology, and geology merely by pointing to anomalies or observations that so far remain unexplained. It is like the knight who attacks the tail of the dragon because he does not have the skill to attack the fire-breathing end.

Thought Questions

1. Who should read this book? Why?
2. Do you find this book threatening? Why? Do you think anyone else will? Why?

CHAPTER 3

# *History of Creationism and Evolutionary Science in the United States*

> In all modern history, interference with science in the supposed interest of religion, no matter how conscientious such interference may have been, has resulted in the direst evils both to religion and science, and invariably; and, on the other hand, all untrammeled scientific investigation, no matter how dangerous to religion some of its stages may have seemed . . ., has invariably resulted in the highest good both of religion and science.
>
> —Andrew Dickson White, cofounder of Cornell University

IN THE BEGINNING, nearly everyone was a creationist, in a manner of speaking. Much about the universe and the earth has the appearance of design, and the only designers we know (humans) are purposeful and intelligent. Lacking any other theory, we find it easy to ascribe the design we see in nature to a deity or, in the case of the Greeks, Romans, and others, deities. To Christian Europeans and Jews, the creator was God; Christian Europe accepted the Hebrew Bible as an accurate account of the creation of the earth. When the Scottish archbishop James Ussher in 1654 examined the chronologies of the Bible beginning with the creation, he was doing cutting-edge scholarship that required a knowledge of ancient history, languages, and cultures, as well as a deep understanding of the text itself. Using the Hebrew Bible as his guide, Ussher concluded that God had created the universe in 4004 B.C.E.

By 1800, however, geological evidence had demonstrated that the earth was far older than a few thousand years (see chapter 15), and the fossil record showed that modern floras and faunas had descended from earlier floras and faunas. Evolution was in the air, so to speak. Charles Darwin's grandfather, Erasmus Darwin, speculated that all warm-blooded animal life had descended from a single ancestor, but he suggested no real mechanism for modification of this ancestral species into the myriad species known in his day. The French naturalist Jean-Baptiste Lamarck in 1802 developed what could be called the first comprehensive theory of evolution.

Lamarck noted that animals are fitted to their environment and thought that a *life force* drove organisms to greater complexity, precisely because of the pressures put on them by the environment. In Lamarck's view, animals developed their characteristics by an act of will and, because of a presumed adaptive force, passed their newly developed characteristics to their descendants. Thus, for example, a blacksmith's descendants will become stronger than he because he acquires powerful muscles from the activity of metalworking and passes them to his descendants. But we cannot breed mice without tails by simply cutting off their tails generation after generation, because, according to Lamarck, there is no adaptive force cutting off the tails. Lamarck's theory, which is now discredited, is often called the *inheritance of acquired characteristics*. It was influential among scientists and natural philosophers of Lamarck's time. Unfortunately, however, the theory could provide no real explanation *why* characteristics might be inherited.

The English naturalist Charles Darwin provided that mechanism after his fateful voyage around the world as the naturalist on the H.M.S. *Beagle* (see chapter 9). Darwin recognized that individual organisms of the same species differ slightly and that the differences were at least partly heritable. He realized that a series of small changes could accumulate over a long time to produce profound changes. Although Darwin did not know how these small changes were transferred from one generation to the next, he saw clearly how beneficial changes could be amplified by competition and selection, rather than by an adaptive force. Another English naturalist and explorer, Alfred Russel Wallace, working primarily in the Amazon and in Indonesia, came to approximately the same conclusion at approximately the same time. Indeed, Wallace's findings inspired Darwin to publish his own theory jointly with Wallace in 1858 and to publish *On the Origin of Species* in 1859.

### The Great Awakenings

Even before Darwin and Wallace's theory of evolution, the discovery of exotic lands and exotic species presented problems for literalist belief in the Bible. How, for example, did animals and plants get from Mount Ararat (after the *Noachian* flood described in the Bible) to the Americas, Australia, and New Zealand? Why do Eurasian and African species differ from their apparently close relatives in the Americas? Why are Australian species strikingly different from species in the rest of the world? Adherents sometimes had to go to great lengths to harmonize their belief in the Bible with new discoveries.

*Higher criticism* is the term generally used for careful, dispassionate efforts to deduce the origin, age, or veracity of various sections of the Bible.

Augustine of Hippo, a Christian church father and philosopher of the fourth to fifth centuries; Desiderius Erasmus, a Dutch philosopher and Catholic theologian of the fifteenth to sixteenth centuries; and Baruch Spinoza, a Dutch-Jewish philosopher and theologian of the seventeenth century all engaged in something like higher criticism, but it originated primarily in the eighteenth century during the *Enlightenment*, a philosophical movement that stressed reason. Higher criticism means, for example, carefully studying alternate versions of the text and alternate narratives within the text, using contemporaneous languages to deduce meanings of obscure passages, and comparing certain narratives of the Bible with narratives from other documents, whether mythical or historical. One intention of higher criticism is to ascertain which parts of the Bible are likely to be historically accurate and which are not. Higher criticism culminated in the twentieth century with the discoveries that the Gospels of the New Testament were not written contemporaneously with the life of Jesus and that the Hebrew Bible consists of several discrete, interwoven threads that tell inconsistent stories.

The Enlightenment was a philosophical and intellectual movement that began in the late 1600s in England, France, and Germany, and stressed the primacy of reason over traditional religious and social ideas. The American Founding Fathers were heavily influenced by the Enlightenment. Interest in religion declined as a result of the Enlightenment and its emphasis on secular learning, and church services in Europe and North America became more intellectual and sterile. The Great Awakening, a religious revival that came to be known as *evangelical*, was in part a reaction against the intellectualism of the Enlightenment and took place in the 1730s and 1740s. During the Great Awakening, sermons and church services became dramatic or emotional rather than intellectual. Preachers stressed sin, punishment, redemption, and the omnipotence of God. The New England preacher Jonathan Edwards argued that faith alone was sufficient for an encounter with God, that reason was neither necessary nor possible. But Edwards and his contemporaries recognized the need not only to confirm their beliefs with reference to scripture, but also to interpret scripture correctly. Although they were *biblical inerrantists*, who believed the Bible to be free of errors and therefore creationists in a sense, they were not *biblical literalists*, and their beliefs were markedly different from those of modern creationists. Even though the Great Awakening stressed religious experience over secular learning, the intellectual ferment of the time spawned several institutions of higher education, including Princeton, Brown, and Rutgers Universities.

In the United States, the Second Great Awakening took place between the early 1800s and the 1840s, and had a progressive philosophy that favored improving society through temperance (abstinence from drinking alcoholic beverages), women's suffrage, and the abolition of slavery. It was thus consistent with nascent evolutionary theories, such as Lamarck's, which saw evolution as a series of progressive improvements. Although most Northern states had abolished slavery by the 1820s, the (mostly Northern) abolitionist movement to abolish slavery in the South gained strength during the Second Great Awakening. Indeed, both the Methodist and Baptist Churches split into Northern and Southern branches in 1845 because of disagreement over slavery. Some evangelical preachers deliberately linked evolution and abolition. Some of the Southern churches may have begun their drift into biblical literalism when they invoked the Bible to justify slavery (see especially Genesis 9:24–27, where Noah curses Canaan, son of Ham).

The initial reception of Darwin's theory in the United States was mixed. When *On the Origin of Species* was published in 1859, it became the subject of debate among leading American scientists, including Louis Agassiz of Harvard, who was arguably the most prominent biologist in the United States. Agassiz accepted that species went extinct but did not accept descent with modification; rather, he insisted that God had created from scratch each new species that appeared in the fossil record. The botanist Asa Gray of Harvard University collaborated with Darwin and was immediately convinced of Darwin's core arguments when he read an early draft of *On the Origin of Species*. Unlike Agassiz, Gray accepted descent with modification, but thought that natural selection was not the only mechanism involved in speciation. Rather, he thought that God must guide the process. Agassiz and Gray thus anticipated some of the arguments of old-Earth creationism and theistic evolution.

Following the American Civil War, many evangelical Christians came to terms with the theory of evolution. They did not give up their biblical inerrantism, but rather interpreted scripture in such a way as to make it consistent with modern science. Thus, they developed the *day-age theory*, in which the "day" of the early chapters of Genesis is taken to mean an age of undefined duration, or various *gap theories* to account for apparent discrepancies between the Bible and evidence from geology and palaeontology (see chapter 6). By the 1880s, however, higher criticism had cast doubt on the inerrancy of the Bible. In particular, the German biblical scholar Julius Wellhausen formulated the *documentary hypothesis*, which argued that the Hebrew Bible was not a single document but rather an edited, or *redacted*, compilation of several documents, all interwoven as wires are woven into a cable.

The Third Great Awakening was in part a response to higher criticism. It began, more or less, in 1886, when the prominent evangelical preacher Dwight L. Moody founded the Moody Bible College. Moody favored biblical literalism and linked evolution with atheism. The Third Great Awakening culminated in the publication of a series of pamphlets called "The Fundamentals" between 1910 and 1915. "The Fundamentals" taught, for example, the inerrancy of scripture, the divinity of Jesus, and the reality of miracles. "The

Fundamentals" themselves were not consistently anti-evolutionary, but some of the essayists opposed natural selection because of its relation to a liberal, naturalistic interpretation of Christianity.

In 1923, George McCready Price, a Seventh-Day Adventist and self-taught geologist, published a book on *flood geology*, in which Price purported to show that geologists' dating of fossils (see chapters 6 and 15) is false and based on circular reasoning. Price's book marks the beginning of modern scientific creationism, but, as a Seventh-Day Adventist, Price was outside the evangelical mainstream and therefore had little influence. His work was updated in 1961 by hydrologist Henry M. Morris and theologian John C. Whitcomb Jr. in their book *The Genesis Flood*, which argued that the earth was only several thousand years old and that the entire fossil record was deposited within a short time by the Noachian flood (see chapter 6). Morris later founded the Institute for Creation Research and became a key spokesman for scientific creationism.

### The Scopes Trial and Its Aftermath

By the 1920s, evolution was to some extent in retreat, at least in the public eye. Many if not most biologists accepted evolution in the sense of descent with modification, but they doubted that natural selection was the most-important mechanism. Intellectuals and religious leaders opposed to eugenics, social Darwinism, and laissez-faire capitalism, not to mention atheism and immorality, blamed the theory of evolution for spreading these ideas. At the same time, partly because of child labor laws, more and more children spent their days in school instead of in the fields and the mills. Public secondary education was booming, and high school students were exposed to biology and evolution. Under pressure from fundamentalists, twenty-three state legislatures considered legislation restricting the teaching of evolution in the public schools. Three states, including Tennessee, actually enacted such legislation.

The fledgling American Civil Liberties Union (ACLU), concerned about restrictions on civil liberties during World War I, actively sought a biology teacher to contest the Tennessee law, the Butler Act, initially on the grounds that it violated teachers' freedom of speech. Hoping to drum up a little business, a Dayton businessman, George Washington Rappleyea, soon arranged for a biology teacher, John T. Scopes, to be prosecuted for violating the Butler Act. In defense of Scopes, the ACLU brought in a legal team including the well-known defense lawyer and agnostic Clarence Darrow. The prosecution imported the progressive-populist politician William Jennings Bryan.

Although the ACLU was initially concerned with free speech, both Darrow and the prosecution intended to show that there was no conflict between the biblical account and evolution. The judge, however, limited the trial to the question of whether Scopes had broken the law. Since Scopes had freely admitted to violating the Butler Act, his conviction was a foregone conclusion. Scopes was fined $100, not an inconsequential sum in 1925. The ACLU appealed the decision with the intention of carrying the case to the Supreme Court of the United States. The Tennessee Supreme Court upheld the constitutionality of the Butler Act but overturned the conviction on the technicality that the judge had imposed the fine, whereas Tennessee law stipulated that the jury must set the penalty. The case was never brought before a federal court.

The Scopes trial lasted for eight days and riveted the country's attention. It was not, however, a victory for evolution. A few states enacted anti-evolution bills after 1925, but, more important, evolution was largely removed from the biology curriculum, as publishers of high-school textbooks shied away from controversy. To some extent, evangelicals were marginalized and stereotyped as backward religious zealots. The opposition to evolution gradually shifted from the urban North to the South, where local school boards adopted anti-evolution policies. The Butler Act was not repealed until 1967.

### The Evolution of Creation Science

Price's flood geology was the only scientific option for creationists when, in 1941, a group of evangelical scientists accepted an invitation from the Moody Bible Institute and founded the American Scientific Affiliation (ASA). Members of the ASA must have at least a bachelor's degree in science or engineering, or the history or philosophy of science. They must subscribe to certain Christian dogmas such as a belief in the Trinity and the "trustworthiness and authority" of the Bible, but also must agree that nature is intelligible and can be described scientifically. In its early years, the ASA favored creationism, but by the late 1950s began to drift toward theistic evolution. Today, many members of the ASA are theistic evolutionists, but the organization itself takes no position on evolution or any other scientific issue. Since 1949 the ASA has published a journal now known as *Perspectives on Science and Christian Faith*.

Evolution remained largely outside the high-school biology curriculum until the 1960s. In 1957, a previously complacent United States watched as the Soviet Union launched the first artificial satellite, *Sputnik*. Besides being a major blow to U.S. national pride, *Sputnik* caused concern that the United States was falling behind in science. The federal government therefore decided to overhaul the nation's science curriculum and improve the quality of high-school science textbooks. Thus, the National Science Foundation (NSF) assembled teams of scientists and teachers to write new textbooks. The Biological Sciences Curriculum Study (BSCS), in particular, found that evolution was largely absent from high-school textbooks. The first BSCS textbook appeared in 1963 and included evolution at its core.

The historian Ronald Numbers notes that most American fundamentalists at the time subscribed to gap theory or day-age theory, which implicitly made certain concessions to scientific thinking (see chapter 6). Most Christians accepted the conventional view of geologic time, and only a relatively few Seventh-Day Adventists adhered to Price's theory of flood geology. In 1961, however, Whitcomb and Morris published *The Genesis Flood*, which modernized Price's flood geology and argued that science should be subservient to scripture. Their argument found a willing ear among conservative fundamentalists; gap theory and day-age theory all but went extinct as flood geology descended, with modification, into scientific creationism and later begat creationist geology, biology, cosmology, and astrophysics. If they could not remove evolution from the science classroom, fundamentalists hoped to use scientific creationism as a wedge to include their religious views alongside evolution (see "Creation Science and the Courts," in this chapter).

In 1963, Morris and other creationists left the ASA and founded the Creation Research Society (CRS). Of the ten founders, six had Ph.D.s in biology or biochemistry, though none had credentials in geology. Although the CRS claims to be a scientific society, it requires its members to subscribe to the following beliefs: (1) The Bible is the word of God, "all its assertions are historically and scientifically true," and the account in Genesis "is a factual presentation of simple historical truths." (2) The basic *types* or *kinds* were created by God during creation week, and biological changes occur only within kinds. (Kind has never been defined properly; sometimes it appears to mean genus and sometimes family.) (3) The Noachian flood was a historical and worldwide event. A great many Americans have come to accept as true some or all of these views.

In the 1970s, Morris edited a textbook, *Scientific Creationism*, and the Creation Research Society published *Biology: A Search for Order in Complexity* in an effort to insert their views into the public school classroom. In 1972, Morris and others founded the Institute for Creation Research (ICR), which proselytizes vigorously in favor of creation science, operates a museum and a publishing house, mails newsletters, and runs workshops. In addition, ICR operates a graduate school that offers degrees in astro-geophysics, biology, geology, and general science. The graduate school has not been accredited by the Western Association of Colleges and Schools, which accredits most other colleges and universities in California, but rather by an accrediting agency for Christian schools, which was founded by Morris himself.

Young-earth creationism is mostly a U.S. phenomenon, but in 1978 an Australian evangelical, Ken Ham, founded the Australian Creation Science Foundation. Ham came to work for the ICR in the United States in 1987 and eventually spun off Answers in Genesis (AIG), which he cofounded with Carl Wieland, a young-earth creationist. AIG founded branches in Australia,

Canada, New Zealand, South Africa, and the United Kingdom, but owing to a schism in 2005 Wieland now heads the non-U.S. branches, which have been renamed Creation Ministries International, while Ham still operates AIG in the United States. In 1986, the Canadian astrophysicist Hugh Ross founded Reasons to Believe, an old-earth creationist organization, but it has never had as much influence as the young-earth creationists.

### Creation Science and the Courts

The religion clause of the First Amendment to the U.S. Constitution reads, "Congress shall make no law respecting an establishment of religion [the establishment clause], or prohibiting the free exercise thereof [the free-exercise clause]." The Fourteenth Amendment extends the protections of the Bill of Rights, which includes the First Amendment, to state governments. Thus, like the federal government, the state may not promote religion, nor may it prohibit the exercise of religion. The legal history of the creationism-evolution controversy in the United States has been shaped by establishment-clause jurisprudence.

In 1963, in a case that did not directly concern creation or evolution, the United States Supreme Court struck down a Pennsylvania law that required daily Bible readings in public schools. In *Abington School District v. Schempp*, the court ruled that a law must exhibit "a secular legislative purpose" and may neither advance nor inhibit religion; it found that the Pennsylvania law did not meet these criteria. Creationists responded with the argument that teaching evolution inhibited religion and fought for *balanced treatment*, that is, teaching creation science whenever evolution is taught in a public school.

In 1965, an Arkansas teacher, Susan Epperson, challenged her state's anti-evolution law on the grounds of free speech. She was joined in her suit by a parent who argued that his child had a constitutional right to study evolution. The case made its way to the United States Supreme Court, which ruled in *Epperson v. Arkansas* that the Arkansas law was unconstitutional because it proscribed teaching evolution on purely religious grounds.

In 1971, the Supreme Court struck down laws that permitted states to reimburse parochial schools for items such as textbooks, supplies, and teachers' salaries. In *Lemon v. Kurtzman*, the court enunciated what has become known as the *three prongs* of the *Lemon test*: If a law concerning religion or a religious institution is to pass constitutional muster, then (1) it must have a secular purpose, (2) it must not primarily advance or inhibit religion, and (3) it must not lead to "excessive government entanglement" with religion. All three prongs of the *Lemon* test must be satisfied if a law is to be constitutional under the establishment clause.

The *Lemon* test figured into the decision of Judge William Overton in *McLean v. Arkansas*. Arkansas passed a balanced-treatment law in 1981. The

Arkansas chapter of the ACLU promptly challenged the law and was joined by a panoply of both religious and scientific organizations. In *McLean*, the plaintiffs argued that the purpose of the law was not scientific but religious, because creation science was inherently religious. They put together a cast of expert witnesses that included a theologian, a palaeontologist, a geneticist, and a philosopher of science. The defense opted not to put Henry Morris of the Creation Research Society on the stand because his obviously religious point of view would have undermined their case. In early 1982, Judge Overton ruled that the Arkansas law violated all three prongs of the *Lemon* test. Specifically, he ruled that the legislators intended to promote a religious viewpoint; that the law's primary effect was to promote a religious view, because creation science is religion and not science; and finally that the law would initiate an excessive government entanglement with religion.

Michael Ruse, a philosopher of science, testified on behalf of the prosecution. Relying on Ruse's testimony, Overton additionally ruled that the "essential characteristics" that make a subject scientific are these: "(1) It is guided by natural law; (2) It has to be explanatory by reference to natural law; (3) It is testable against the empirical world; (4) Its conclusions are tentative, i.e.[,] are not necessarily the final word; and (5) It is falsifiable." He found that scientific creationism fails on all counts. His ruling was not appealed.

In 1982, Louisiana passed its own balanced-treatment act, which was similar to the Arkansas statute, but its framers hoped they had sanitized it of any taint of religion. The ACLU challenged the law, and the Federal District Court ruled the law unconstitutional because it advanced a religious view, inasmuch as it required creation science to be taught in the public schools if evolution was taught there. Ultimately, in 1987, in *Edwards v. Aguillard*, the United States Supreme Court struck down the Louisiana act as unconstitutional, writing, "The preeminent purpose of the Louisiana Legislature was clearly to advance the religious viewpoint that a supernatural being created humankind."

*Edwards v. Aguillard* thus killed the balanced-treatment approach once and for all. Creationists had to try a different tactic to get around the Constitution, and they found comfort in one line in the majority opinion, which said that "teaching a variety of scientific theories about the origins of humankind to schoolchildren might be validly done with the clear secular intent of enhancing the effectiveness of science instruction." They found further comfort when Justice Antonin Scalia wrote in dissent, "The people of Louisiana, including those who are Christian fundamentalists, are quite entitled, as a secular matter, to have whatever scientific evidence there may be against evolution presented in their schools, just as Mr. Scopes was entitled to present whatever scientific evidence there was for it." These two sentences foreshadowed two new stratagems: relabeling creationism as "intelligent design" and campaigning for "critical thinking." The *Edwards* decision was followed by a handful of other court decisions to the effect that a school district may require a teacher to teach biology and prohibit a teacher from teaching creationism.

In 1994, the Tangipahoa, Louisiana, Parish Board of Education adopted a policy that "the [biology] lesson to be presented, regarding the origin of life and matter, is known as the Scientific Theory of Evolution and should be presented to inform students of the scientific concept and not intended to influence or dissuade the Biblical version of Creation or any other concept." The board further recognized "that it is the basic right and privilege of each student to form his/her own opinion or maintain beliefs taught by parents on this very important matter of the origin of life and matter. Students are urged to exercise critical thinking and gather all information possible and closely examine each alternative toward forming an opinion." In 1997, in *Freiler v. Tangipahoa Parish Board of Education*, the Federal District Court, in addressing this statement, ruled that the Board of Education was endorsing religion by implicitly conveying the message that evolution is a religion that may contradict the religious beliefs of the students. The decision was upheld in 1999 by the Court of Appeals, and the Supreme Court declined to review the case.

In 2002, the Cobb County, Georgia, school district mandated that biology textbooks be supplemented with a sticker that read, "This textbook contains material on evolution. Evolution is a theory, not a fact, regarding the origin of living things. This material should be approached with an open mind, studied carefully, and critically considered." In Federal District Court, Judge Clarence Cooper found in *Selman v. Cobb County School District* that the sticker violated the second prong of the *Lemon* test: its primary effect was to advance religion. Owing to concern about the evidence submitted during the trial, the Court of Appeals vacated the decision and sent it back to the District Court. The case was eventually settled when the school district agreed not to replace the stickers (which had been removed after the trial court's decision), not to take any further action to deprecate evolution, and to pay over $160,000 toward the plaintiffs' attorneys' fees.

### Intelligent-Design Creationism

In 1989, the little-known Foundation for Thought and Ethics published a supplementary biology textbook, *Of Pandas and People*. The book uses mostly standard young-earth creationist arguments, stripped of their overtly religious language, to try to refute the theory of evolution. *Of Pandas and People* was the first systematic presentation of intelligent-design creationism and was in preparation when the Supreme Court handed down the *Edwards* decision. The book later morphed into the slick volume *The Design of Life* (2007).

Intelligent-design creationism began in earnest with the publication of *Darwin on Trial*, by Phillip Johnson, in 1991. Johnson, then a law professor at the University of California at Berkeley, repeated many of the arguments of young-earth creationism but stressed that science uses only naturalistic explanations and arbitrarily—indeed, dogmatically—rules out the supernatural (see chapter 4). Johnson thus clearly states his intention to defeat naturalism. In 1996, he helped found the Center for Science and Culture (initially the Center for the Renewal of Science and Culture) at the Discovery Institute, a well-funded right-wing think tank. In 1998, the Discovery Institute drafted the Wedge Document, which, though labeled top secret, was shortly leaked on the Internet and later authenticated by Barbara Forrest, a philosophy professor at Southeastern Louisiana University. The Wedge Document describes a campaign to destroy scientific materialism and its supposed destructiveness to morality and culture, and to replace it with a theistic understanding of nature. The thin edge of the Wedge was Johnson's *Darwin on Trial*; Johnson's next book, *The Wedge of Truth: Splitting the Foundations of Naturalism*, appeared in 2000.

The Wedge strategy included a five-year plan to publish thirty books and one hundred technical papers, as well as to "pursue possible legal assistance" to force intelligent-design creationism into the public schools. After more than a decade, various fellows of the Discovery Institute have published a number of books, all of which were dismissed by reputable scientists, but they have initiated no original scientific research that supports the concept of intelligent design. Although its religious orientation is explicit, the long-term plan outlined in the Wedge Document also displays the Discovery Institute's political agenda very clearly. In ten years, the Wedge strategy was to be extended to ethics, politics, theology, the humanities, and the arts. The ultimate goal of the Discovery Institute is to "overthrow" materialism and "renew" American culture to reflect right-wing Christian values.

### Attacks on Evolution

In 1996, Michael Behe, a biology professor at Lehigh University, published the book *Darwin's Black Box*. Behe argued that certain biochemical structures, such as the bacterial flagellum and the human blood-clotting system, were *irreducibly complex* and could not have evolved by natural selection (see chapter 8). His arguments have been refuted many times, and he has never satisfactorily responded to his critics. William Dembski, who has Ph.D.s in both mathematics and philosophy, developed a mathematical formalism that both subsumed Behe's argument and purported to put it on a sound mathematical basis. Using a smokescreen of complex terminology, Dembski advanced an oversimplified probabilistic argument and concluded, in effect, that an organism could not have been assembled in one fell swoop. Later, Dembski misused the no-free-lunch theorems (see chapter 8) in an effort to show that evolution was impossible.

In 2002, Jonathan Wells, a fellow at the Discovery Institute, published *Icons of Evolution: Science or Myth? Why Much of What We Teach about Evolution Is Wrong.* Wells entered graduate school and earned a Ph.D. in religious studies from Yale University with the specific intention to "destroy Darwinism"; he later earned a second Ph.D. in molecular and cellular biology. His studies at Yale were funded by the Unification Church. In *Icons*, Wells identifies ten case studies that he claims are commonly used as exemplars in high-school biology textbooks to illustrate the theory of evolution, labels them icons, and purports to show that they are used not as illustrations but as evidence.

Eugenie Scott of the National Center for Science Education has examined the textbooks that Wells cites and found that most of them either do not include Wells's exemplars or do not use them in the manner that Wells describes. Wells's book has nevertheless been used to pressure school boards to drop certain biology textbooks; anecdotal evidence suggests that excellent exemplars of evolution, such as the peppered moth (see chapter 1), are being removed from textbooks because of disinformation promulgated by creationists like Wells.

### Teach the Controversy

With the publication of the books by Johnson, Behe, Dembski, and Wells, the Discovery Institute thus manufactured a controversy and demanded that it be taught under the rubric of critical thinking: If evolution is taught, then intelligent-design creationism must also be taught.

In 1999, the Kansas Board of Education removed all references to evolution from that state's science standards. Science organizations warned of a serious loss in the quality of science education. Two years later, a new Board of Education reinstated evolution. The Board changed hands again and in 2005 adopted new standards that portrayed evolution as a theory in crisis. The new standards mirrored the Discovery Institute's Critical Analysis of Evolution strategy. In response, both the National Academy of Sciences and the National Science Teachers Association withdrew permission to use their copyrighted materials in the standards. Earlier in the year, the Board had held a series of hearings that was attended by most of the prominent figures in intelligent-design creationism but was boycotted by mainstream scientists. In November 2005, the Kansas Board of Education changed the science standards to state that evolution is a theory, not a fact; to consider supernatural explanations as part of science; and to require students to be informed of the purported controversy. In 2006, the composition of the Board changed again, and in 2007, the original 2005 standards were reinstated.

In 2002, meanwhile, the Ohio State Board of Education hosted a nonbinding discussion of evolution involving Jonathan Wells and Stephen Meyer of the Discovery Institute for the creationist side and biologist Kenneth Miller of Brown University and physicist Lawrence Krauss of Case Western Reserve University on the side of science. The board adopted a set of science standards that included a clause concerning critical analysis of evolution; an apparently innocuous clause, it singled out evolution among all the sciences and provided a handhold for anti-evolutionism. In 2004, the board approved grade-10 lesson plans that were based on the Discovery Institute's Critical Analysis of Evolution and purported to provide an opportunity for critical analysis of Darwin's theory. The board eventually dropped the lesson plans and the standard, partly in response to a contemporaneous school board case, to which we now turn.

In Dover, Pennsylvania, several members of the school board expressed concern about the teaching of evolution. The school board voted to amend its policy by adding the following language: "Students will be made aware of the gaps/problems in Darwin's theory and of other theories of evolution including, but not limited to, intelligent design. Note: Origins of life is [sic] not taught." The Thomas More Law Center, a right-wing Christian organization, offered to represent the board if its policy was challenged. To implement that policy, the board required teachers to read to their biology classes a statement, "Because Darwin's Theory [*sic*] is a theory, it is still being tested as new evidence is discovered. The Theory [*sic*] is not a fact. Gaps in the Theory [*sic*] exist for which there is no evidence. . . . Intelligent design is an explanation of the origin of life that differs from Darwin's view. The reference book, *Of Pandas and People*[,] is available for students to see if they would like to explore this view in an effort to gain an understanding of what intelligent design actually involves. . . . [S]tudents are encouraged to keep an open mind. The school leaves the discussion of the origins of life to individual students and their families." District science teachers unanimously refused to read the statement in class; it was read instead by school administrators.

Lawyers from the ACLU, Americans United for the Separation of Church and State, and the law firm Pepper Hamilton filed suit on behalf of concerned parents and teachers in December 2004. The case, *Kitzmiller v. Dover Area School District*, was heard in 2005 by Judge John E. Jones III, an appointee of George W. Bush. The Foundation for Thought and Ethics, the publisher of the book *Of Pandas and People*, petitioned to intervene as a codefendant, citing potential losses if the plaintiffs won the suit, but Jones rejected their petition because it was not timely and because they had not shown that the existing defendants would not adequately represent their interests. Three expert witnesses for the defense ultimately declined to testify in a dispute over legal representation.

The plaintiffs argued that several school board members had religious motivations and therefore that the board policy was unconstitutional. Additionally, Barbara Forrest, the philosophy professor who had authenticated the Wedge Document, showed that early drafts of the book *Of Pandas and People* used the words *creationist* or *creationism* approximately one hundred times. A later draft, drawn up after the 1987 *Edwards* decision, replaced all instances of "creationist" with "design proponent." Indeed, one apparently hasty cut-and-paste operation morphed "creationists" into the transitional form "cdesignproponentsists." Thus did creationism evolve into intelligent design.

Judge Jones found for the plaintiffs. He ruled that intelligent design is "a religious view, a mere re-labeling of creationism, and not a scientific theory." He gave short shrift to the argument that the school board's policy was not advocating the teaching of intelligent-design creationism but merely making students aware of it. Finally, Jones ruled that "the secular purposes claimed by the board amount to a pretext for the board's real purpose, which was to promote religion in the public school classroom, in violation of the Establishment Clause." In the November elections before the verdict, a school board favoring evolution was elected, so the decision was not appealed. The school district, however, was left with a $2 million debt for attorney's fees and court costs. The plaintiffs' legal team settled for $1 million, plus one dollar for each of the eleven plaintiffs.

Creationists may be expected to change their tactics again in the wake of the *Kitzmiller* decision. As we go to press, a half-dozen or so state boards of education are considering policies that use the rubric of *academic freedom* to mandate teaching the "strengths and weaknesses" of evolution, but not of any other subjects. The proponents of such mandates deny that their motivation is religious. In 2008, Governor Bobby Jindal of Louisiana, a biology major from Brown University and a Rhodes scholar, signed a law that allowed the State Board of Elementary and Secondary Education to "assist teachers, principals, and other school administrators to create and foster an environment within public elementary and secondary schools that promotes critical thinking skills, logical analysis, and open and objective discussion of scientific theories being studied including, but not limited to, evolution, the origins of life, global warming, and human cloning." The bill further allows a teacher to "use supplemental textbooks and other instructional materials to help students understand, analyze, critique, and review scientific theories in an objective manner," at the discretion of the local school board. Perhaps in a case of protesting too much, the law affirms, "This Section shall not be construed to promote any religious doctrine, promote discrimination for or against a particular set of religious beliefs, or promote discrimination for or against religion or nonreligion."

## Conclusion

Until the development of a truly scientific theory of evolution in the nineteenth century, most scholars and theologians probably thought that the plants and animals they saw around them had been designed by the Creator. Darwin and Wallace's theory made such an assumption unnecessary. In the United States, many religious believers came to terms with the new theory of evolution, for example, by developing gap and day-age theories. In the early 1900s, however, many evangelical Christians began to drift toward fundamentalism and became suspicious of evolution. Evolution was not a major factor in their thinking after the Scopes trial, if only because it had largely disappeared from the high-school curriculum. After the launch of *Sputnik* in 1957, when biology curriculums began to stress evolution, opposition grew, and creationists developed, in turn, flood geology, scientific creationism, and intelligent-design creationism. State and local school boards tried many stratagems to discredit evolution or to inject some form of creationism into the high-school science curriculum. They were prevented from doing so primarily by strong legal action on the part of evolution supporters.

In succeeding chapters, we will show that all the substantive claims of creationism, including, in particular, intelligent-design creationism, are false. To that end, we will first investigate how science works and then how pseudoscience works. Only after we understand both science and pseudoscience can we hope to understand creationism.

## Thought Questions

1. Do most people change their minds when presented with new evidence? Or do they deny the evidence or try to explain it away? Is this a natural reaction? What is the best way to overcome this obstacle? What do you do?
2. What if the weight of evidence or the weight of informed opinion clearly supports one position over another? Must the second opinion be given equal time with the first? What if a majority of the population supports the second opinion? Explain your answers.
3. Are high-school students competent to evaluate the arguments in a purely scientific controversy? Laypersons in general? Who is competent?

PART TWO

# *How Science Works (and Creationism Doesn't)*

# Chapter 4

## *How Science Works*

> If it disagrees with experiment it is wrong. In that simple statement is the key to science. It does not make any difference how beautiful your guess is. It does not make any difference how smart you are, who made the guess, or what his name is—if it disagrees with experiment it is wrong. That is all there is to it.
>
> —Richard Feynman, Nobel Laureate in physics, 1965

We will argue in this book that intelligent-design creationism (ID creationism) is not science but pseudoscience. That is, it uses the terminology of science, seemingly applies the tools of science, but in fact only masquerades as science. Before we can establish what is a pseudoscience and what is not, however, we need to examine how science works.

What do scientists do when they want to find a new law? We'll let you in on a little secret: They guess at it.

But they don't stop there. They don't presume that their guess is right and proceed to the next guess. To the contrary, they test their guess by figuring out some consequences of their guess and comparing them with nature. If the comparison with nature fails, then the guess is wrong, no matter how appealing it was, no matter how smart the guesser was, no matter how firmly he or she believed it. If, on the contrary, the guess seems to work, then scientists will continue to look for more evidence and make more guesses.

A body of knowledge and inference that explains a vast number of related observations is called a *theory*. In science, a theory is not a guess or a hunch but a complete system of guesses and consequences that supports or explains experimental results or observations. Theories stand or fall on the accuracy of their predictions. In what follows, we'll examine the manner in which theories are developed and show how science progresses from guesses (or hypotheses) to accepted facts.

### Newton's Law of Gravity

In the seventeenth century, the English mathematician and natural philosopher Isaac Newton guessed that the gravitational force between two bodies would depend on the distance between them (technically, that force is proportional to the inverse square of the distance between them). He already knew from calculations performed by the German astronomer Johannes Kepler that the orbits of the planets were not perfect circles but rather types of ovals called ellipses. Newton used his theory to calculate the expected orbits of the planets and found that the theory correctly derived elliptical orbits.

Kepler's observations and Newton's theory were thus in good agreement. Does that mean that Newton's theory is right? No; it means only that the theory is not obviously wrong. Newton and others performed countless additional calculations over two hundred years and found that Newton's theory of gravitation agreed with experimental observation as closely as anyone could tell.

Newton and other astronomers thought that Newton had hit on a universal truth. Today we would conclude that Newton's theory is only tentatively correct or, more precisely, that it is correct within certain limits.

Newton's theory is a scientific theory in part because it is *testable*. To be testable means that the theory has inspired some predictions and that those predictions can be compared with nature. If the predictions had not agreed with observations, then the theory would have been rejected, and someone else would have had to make a different guess. That guess would then have been tested, and possibly that person's theory would have been accepted.

Sometimes experimenters lead theorists and make discoveries that theorists had not predicted. It turns out that the orbits of the planets are not precisely ellipses; in fact, they do not quite close on themselves. Why? Mostly because the gravity of the giant planets, Jupiter and Saturn, slightly distorts the orbits of the other planets. Astronomers have used Newton's theory of gravity to calculate the effects of all the planets on the orbit of Mercury and found that they could account for its path very well, except for a small discrepancy. Even when corrected for the gravitational attraction of the planets, the orbit of Mercury is not precisely a stationary ellipse but is more closely described as an ellipse whose axis rotates at the rate of approximately 43 arcseconds (about 1/100 degree) per century, as shown in figure 1. The rotation of the axis is called the *precession* of the axis and cannot be explained by Newton's theory.

Newton's theory thus failed as a universal truth—it could not wholly account for all of the difference between Mercury's orbit and a stable ellipse. But because the theory had been so successful at predicting everything else, it was not rejected out of hand; instead the discrepancy was temporarily ignored. It suggested, however, that Newton's theory was not the last word, that it might some day be replaced by a more precise theory.

That day came in 1915, when the German-Jewish physicist Albert Einstein proposed his general theory of relativity. Relativity is a theory of gravity that is far too complex to describe here. It encompasses Newton's theory but provides more accurate calculations in strong gravitational fields.

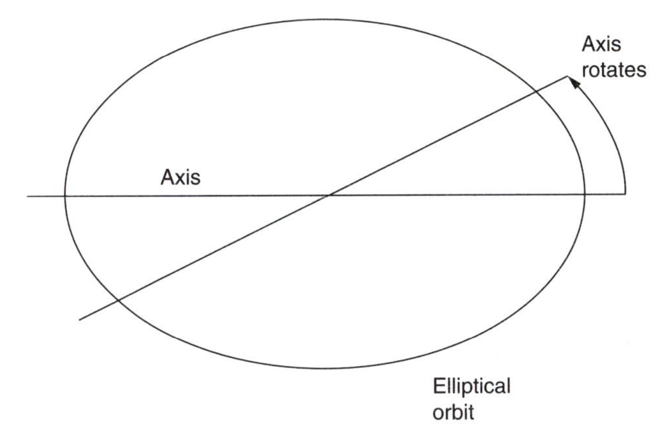


1. The orbit of Mercury is approximately an ellipse. Its axis rotates approximately 40 arcseconds per century, an effect explained by general relativity but not by Newtonian mechanics.

General relativity successfully described the precession of the axis of the orbit of Mercury, where Newton's theory had failed.

Is general relativity correct then? It has passed every test we have so far devised, but we know it cannot be an ultimate theory because it cannot be made consistent with an equally successful theory, quantum mechanics. We are still looking for the theory that combines general relativity with quantum mechanics.

### Theory and Experiment

There is a constant interplay between theory and experiment. To be useful, a theory must make specific predictions. Those predictions must be compared with experiment or observation, with what nature tells us. Eventually, however, an experiment or observation will not agree with some hypothesis, and theorists may have to develop another theory—beginning with another guess—which experimentalists will test again and again until the theory is accepted or rejected. If it is rejected, theorists will have to turn to yet another guess.

The interplay between hypothesis and experiment or observation tells us that we can't just propose a theory and expect it to be right; we have to test it. And we have to test it over and over to make sure it is right. In a way, we have to try our best to *disprove* our own theory, lest we get gulled into accepting a wrong theory. Even if our new theory is successful, that is, generally accepted and useful, it may eventually be overthrown by another, better theory.

It is very rare, however, for a successful theory in a mature science like physics, chemistry, geology, or evolutionary biology to be completely overthrown. Almost always, it is incorporated in some way into a better, more general theory. Specifically, a new theory must yield the same results as an older theory in the domain in which the older theory yielded results that are in agreement with experiment. In short, a successful theory is almost never thrown out; rather, it is subsumed into a newer, more general theory. Thus, for example, the theory of fluid dynamics proposed in the eighteenth century by the Swiss mathematician Daniel Bernoulli was not discarded when scientists discovered that fluids are composed of molecules. Rather, the molecular theory was developed in such a way that it included Bernoulli's theory and indeed explained it.

What then makes a good scientific theory? It has to make precise predictions, such as Newton's prediction that the planetary orbits are elliptical. It has to be fruitful. That is, it has to suggest experiments or observations that will test the theory or else lead to new theories. And it has to be testable (which is a shorthand way of saying that its predictions have to be testable). A theory that is not testable is not a valid theory, scientific or otherwise. As we will see, pseudoscientists propose a variety of theories that cannot be tested and so are useless. But first, let us look more closely at how competent scientists test their theories by examining briefly two historical examples, one successful and one not: continental drift (successful) and polywater (not).

### METHODOLOGICAL NATURALISM

Science in general does not so much reject the supernatural as ignore it. Science cannot incorporate the supernatural into its methodology; it could not function if it had to contend with occasional violations of natural law, but must instead assume that nature is regular and repeatable. This assumption is known as *methodological naturalism*. Methodological naturalism is not a choice; science cannot function if it is not *empirical*, that is, ultimately based on experiment and observation. Science necessarily restricts itself to purely naturalistic explanations.

Phillip Johnson, whom we met in chapter 3, makes it appear as if science's use of methodological naturalism is merely a matter of choice and, indeed, is wholly unjustified. Johnson and other intelligent-design creationists conflate methodological naturalism with *philosophical naturalism* (also called *ontological naturalism*; an *ontology* is a theory of existence). Philosophical naturalism, however, is more extensive than methodological naturalism in that it presumes that there are no causes other than natural causes; methodological naturalism does not necessarily imply philosophical naturalism and therefore has no theological implications.

### Continental Drift

A German meteorologist, Alfred Wegener, proposed the theory of *continental drift* in 1912, when geologists considered the earth's surface solid and unchangeable. Wegener, however, had noticed that the coastlines of Western Europe and Africa were a mirror image of the coastlines of the Americas and hypothesized—guessed—that the continents had once been joined but later drifted apart.

Wegener did not merely declare that his theory was obvious. Instead, he amassed a great deal of evidence to support his hypothesis. He showed that the elevations of features on the earth's surface were not what we might expect if the earth had simply cooled and solidified. He noted that the longitude of a certain island near Greenland had apparently shifted about one kilometer westward between 1823 and 1907. And, finally, he pointed to the unsolved problem that fossils of related species were found, for example, in South America and Africa but not in Europe or North America, precisely as we might expect if Africa and South America had once been joined.

In spite of the wealth of evidence in its favor, Wegener's theory was not generally accepted until the 1960s, with the discovery of firm physical evidence of material welling up from the ocean floor. This evidence was considered definitive, and Wegener's theory was then accepted relatively rapidly.

We could argue that geologists should have been faster to accept Wegener's theory, that their reluctance establishes the claim that scientific acceptance is arbitrary and depends on the prevailing scientific dogma. In defense of the geologists, however, we would counter that there are a lot of crackpot theories, and sometimes it is hard to recognize an original but bizarre contribution. Further, there were serious technical objections to Wegener's theory. In particular, the earth was generally thought to be a rigid body. No one had a plausible idea where sufficient force to move the continents could have originated. Calculations suggested that drift was not possible. In any event, when the evidence of material welling up from the ocean floor became known (see chapter 15), the theory was adopted almost overnight—on the weight of evidence. Indeed, once that crucial bit of evidence was recorded, the revolution took less than a decade.

### Polywater

On a different note, consider the case of *polywater*. This was supposedly a form of water that had different properties from normal liquid water: a different boiling point, a different freezing point, and a different viscosity (resistance to flow). It was first reported in the Soviet Union and was introduced to the West in the mid-1960s by Boris Deryagin of the Institute of Physical Chemistry in Moscow. Deryagin was a respected Soviet scientist and the

IS EVOLUTION "JUST A THEORY"?

We won't answer that question just yet, but we want to make clear that "theory" in science is not equivalent to "speculation" or even "hypothesis." According to the National Academy of Sciences, "Theory: In science, a well-substantiated explanation of some aspect of the natural world that can incorporate facts, laws, inferences, and tested hypotheses."

Paul Strode teaches the distinctions among these terms by asking his students to write scientific hypothesis statements of the form, "*If* hypothesis X is true, *and* I perform method Y, *then* I predict Z as a specific, measurable outcome." He calls this format the *research hypothesis.* He uses that format because it describes how humans actually think about a problem or an observation. For example, he begins by describing a person who is preparing to go camping. The camper pulls her box full of camping gear out of storage and in checking through all of her supplies observes that her big camping flashlight doesn't work. At this point she wonders, "Why doesn't my flashlight work?" Immediately, she comes up with two possible explanations: (1) the batteries are dead, and (2) the bulb is burned out. She has just formulated two hypotheses for why the flashlight does not work. She then designs an experiment to test her first hypothesis by replacing the old batteries with new ones and predicts that the flashlight will then work. She has perhaps unknowingly, but naturally, developed a research hypothesis: *if* the dead batteries hypothesis is true, *and* I replace the old batteries with new ones, *then* I predict that the flashlight will work. If the prediction is borne out, it supports the original hypothesis. The prediction cannot prove the hypothesis, however, because simply replacing the batteries could have repaired an electrical connection that had been broken when she first picked up the flashlight.

Strode describes a specific situation and then asks his students, "What do you predict will happen in this situation?" The students make a prediction, and he writes it on the bottom of the board at the front of the room. He then asks them, "Well, why did you make that prediction?" They explain (usually beginning their explanation with "Because . . .") what knowledge allowed them to state the hypothesis and intelligently make the prediction. He writes this statement up near the top of the board. He finally asks them to explain how they are going to use their prediction to test their hypothesis and writes their answer on the board between the other two statements. He then erases "Because" and replaces it with "If," connects the first and second statements with the word "and," and the second and third statements with "then." Magically, but with a little direction, the students have developed a tentative, scientific hypothesis, complete with a method to test it.

Many teachers and therefore their students mistakenly confuse hypothesis with prediction. Here are three examples of how students are wrongly taught to think about and write hypotheses that guide experimentation, followed by a correctly written, testable hypothesis. The observation driving the following experiment is that when our hands are cold, they seem to function less well. The experiment involves students breaking as many toothpicks as possible with one hand for one minute when their hands are at body temperature, and then after five minutes of soaking their hands in ice water.

1. I can break more toothpicks with my warm hands than I can when my hands are cold. (No. This is simply a prediction, not a hypothesis in the scientific sense.)
2. *If* I break toothpicks for one minute with my warm hands and then for one minute with my cold hands, *then* I will break more toothpicks with my warm hands. (No. This is a method followed by a prediction—there is no apparent reason for doing this experiment. What explanation is being tested? This may be the most common wrong way students and their teachers write hypotheses.)
3. *If* I break toothpicks for one minute with warm hands and then for one minute with my hands after soaking them in ice water, *then* I will break more toothpicks with warm hands *because* low temperatures suppress muscle contractions. (Almost. But this form puts the hypothesis being tested, that cold suppresses muscle contractions, at the end of the statement, in the conclusion, rather than in the beginning where the hypothesis belongs. Also, the use of the word "because" suggests certainty and removes the necessarily tentative nature of the hypothesis.)
4. *If* low temperatures suppress muscle contractions, *and* I break toothpicks for one minute with warm hands and then for one minute with my hands after soaking them in ice water, *then* I will break more toothpicks with warm hands. (Yes. This begins with the hypothesis that low temperatures suppress muscle contractions, and beginning with the word "if" makes the hypothesis tentative. This form also includes how this hypothesis will be tested, and ends with a specific, measurable, predicted outcome of the experiment.)

Students find that they do in fact break fewer toothpicks with their cold hands than with their warm hands. In a typical class of 30 students, the mean number of toothpicks broken with warm hands is approximately 36, and the mean number of toothpicks broken with cold hands is approximately 22. The results therefore support (but do not prove) the hypothesis that low temperatures suppress muscle contractions.

The theory of evolution encompasses countless examples of testable hypotheses and specific, measurable predictions; the vast majority of these have held up to considerable testing and predictions or, when necessary, have been replaced by better hypotheses and related predictions. Obviously, the theory is not perfect, and we still have much to explain, but overall the theory is so successful that it is almost impossible to understand biology without it. Yes, evolution is a theory, but, no, it is not "just" a theory. To the contrary, it is one of the most successful and comprehensive theories in all of science.

winner of a prize that was roughly equivalent to the National Medal of Science in the United States.

Solids and liquids are held together by electrical forces between molecules. The molecules in a liquid are not bound together as tightly as those in a solid but rather are comparatively free to move around; this freedom is what gives a liquid its fluidity. Polywater seemed to behave differently than ordinary water, and scientists at the U.S. National Bureau of Standards (now the National Institute of Standards and Technology) hypothesized that it was a form of water in which the molecules are bound together in long chains, or polymers, like those in a plastic. They and others developed mathematical theories designed to calculate the properties of a plastic (or polymerized) form of water.

Polywater could be prepared only in fine silica or glass capillaries; attempts to produce it in large quantities failed. Silica (silicon dioxide) is one of the main components of common glasses and is slightly soluble in water. When other experimenters found impurities, including silica, in their samples of polywater, Deryagin explained that it was necessary to keep everything extremely clean and faulted other researchers for not maintaining sufficient cleanliness. Eventually, however, the evidence began to show that the unusual properties were the result of silica dissolved in the water. Within a comparatively short time, Deryagin recanted and agreed that polywater was not a new form of water, but rather the result of contamination by dissolved silica.

### AUXILIARY HYPOTHESIS

Sometimes a scientific theory seems to go wrong. Everything works up to a point, and then an experiment, for example, contradicts accepted knowledge. Scientists have two choices in such cases: abandon the old theory and find a new one, or find a way to modify or expand the old theory.

Physicists believe strongly in the "law" of conservation of energy. Over the years, physicists realized that there were many forms of energy and postulated that energy can neither be created nor destroyed; rather it is *conserved*, or transformed from one form to another. The *law of conservation of energy* is thought to be universally true, and indeed no violation of this law has ever been found.

When the Austrian physicist Wolfgang Pauli examined a certain nuclear reaction, though, he found that energy appeared not to have been conserved. Reluctant to give up the idea of conservation of energy, Pauli developed an *auxiliary hypothesis*: He speculated that the missing energy was carried off by a new particle, which we now call a *neutrino*. Pauli didn't stop there, however. He developed a mathematical model to predict certain properties of neutrinos, so that their existence could be verified. Neutrinos were found twenty-five years later and are now well-established, even if still hard to detect. Had neutrinos not been detected or had they been shown to have the "wrong" properties—properties that differed from Pauli's predictions—then Pauli's theory would have failed, and he would have had to develop another auxiliary hypothesis or look for another theory entirely. Merely stating the auxiliary hypothesis was not enough: Pauli had to deduce consequences of the hypothesis, and someone had to test them. Deryagin's claim that other experimenters had not kept their apparatuses sufficiently clean, incidentally, was also an auxiliary hypothesis, but it was invalid, and Deryagin shortly discarded it.

### Is Evolution Testable?

The theory of evolution is indeed testable, and it has been tested over and over, with stunning success, as we will show in later chapters. That is not to say that there are no unsolved problems or inconsistencies; as we have seen, even a successful theory will have inconsistencies, but these are not necessarily enough to defeat a theory.

When people ask whether evolution or any other science is testable, they are asking whether any single experiment or observation is so critical that it will conclusively count against it. That is, they ask whether evolution can be *falsified*. That is not exactly the right question. The right question is, "Does the theory of evolution make testable or falsifiable predictions?" Predictions can be tested, theories cannot.

Einstein said, with some exaggeration, "No amount of experimentation can ever prove me right; a single experiment can prove me wrong." Certainly, Sir Karl Popper, an influential philosopher of science, was mistaken when he said that scientists try as hard as they can to falsify their own theories. To the contrary, like anyone else, scientists probably try their hardest to confirm those theories. It is thus more nearly correct to say that a scientific theory has to be falsifiable in principle—that there has to be some observation or experiment that could count against it or seriously damage it. Even then, as we have seen, a scientist is apt to develop an auxiliary hypothesis to explain an unexpected result. That hypothesis itself has to be falsifiable, however, or it is not scientific.

Many people do not understand falsifiability. They think it means that we have to specify an experiment or observation that will actually falsify a theory. A successful theory, however, has already survived many attempts at falsification.

### POPPER ON EVOLUTION

Creationists often quote Sir Karl Popper to the effect that evolution is tautological and therefore not scientific. What Popper actually said was that natural selection, not evolution, was not falsifiable but rather was a "metaphysical research programme" that was not tautological but "almost tautological." By metaphysical, Popper meant that natural selection was not falsifiable according to his strict standard, not that it was not useful. To Popper, a metaphysical research program was the set of rules or assumptions, sometimes held unconsciously, that guide scientists in their activities. Many of these assumptions are unprovable but are nonetheless important. Determinism, for example, was a metaphysical assumption that most physicists abandoned after the development of the quantum theory. Indeed, empiricism, the concept that the nature of the universe can be learned by observation and experiment, is a metaphysical assumption.

Thus, Popper was not being negative when he described natural selection as a metaphysical research program, and he argued that natural selection provided a framework for developing a falsifiable scientific theory. Indeed, he called natural selection a theory and claimed that it was "invaluable. I do not see how, without it, our knowledge could have grown as it has done since Darwin. In trying to explain experiments with bacteria which become adapted to, say, penicillin, it is quite clear that we are greatly helped by the theory of natural selection. Although it is metaphysical, it sheds much light upon very concrete and very practical researches."

Whatever his initial position on natural selection, Popper in 1978 explained that he had been trying to explain how natural selection could be untestable and still scientifically interesting. His solution had been to define it as a metaphysical research program that guides our thinking. Later he admitted that natural selection was difficult but not impossible to test and in so many words recanted: "I still believe that natural selection works this way as a research programme. Nevertheless, I have changed my mind about the testability and logical status of the theory of natural selection; and I am glad to have an opportunity to make a recantation. My recantation may, I hope, contribute a little to the understanding of the status of natural selection." His recantation has, unfortunately, gone unnoticed by most creationists.

It is falsifiable in principle precisely because it has been subjected to rigorous testing. We will see shortly how pseudoscientists evade the requirement that their theories be falsifiable by presenting auxiliary hypotheses that are vague or unfalsifiable and that allow the pseudoscientist to explain away any observation that might cast doubt on a theory.

How is evolution testable? The influential biologist J.B.S. Haldane was once asked what would constitute evidence against evolution. He famously answered, "A fossil rabbit in the Cambrian." In other words, any fossil that was out of place, in the wrong geological stratum (see chapter 6, "Why Creationism Fails"), would count as serious evidence *against* evolution, unless its presence there could be otherwise explained. Indeed, evolution has to be relatively gradual, so if a platypus gave birth to a rabbit, Cambrian or not, evolution would be falsified. Similarly, if we once, anywhere on earth, discovered an organism that had appeared out of nowhere—*spontaneous generation*—then evolution will have been cast into serious doubt because we will then realize that not all present life has descended from earlier life.

The tree of life, or *phylogeny*, is a model that shows how species are related and how different species have descended from a common ancestor. Naturalists such as botanists and zoologists have developed a phylogeny based on the shared characteristics among species and on geographical relationships between species. Palaeontologists have similarly developed a phylogeny based on the fossil record. Molecular biologists have developed a phylogeny based on genetics. These three independent phylogenies do not agree perfectly with one another, but they show broad general agreement (see chapter 11). If they had differed greatly and in unexplained ways from each other, then the theory of evolution would have been in serious trouble, if not outright falsified. In fact, the three phylogenies agree well enough to support and sometimes correct each other. Further, the closer the relation between two species according to shared characteristics, the more alike are their genomes; if their genomes varied randomly, for example, then again evolution would be in trouble. To the contrary, the predictions of evolutionary biologists have been borne out time and again.

Additionally, evolution requires an old earth. In Darwin's time, the antiquity of the earth was by no means an established fact, and a young earth would have conclusively refuted Darwin's theory. Indeed, the theory of evolution predicted an old earth, and ultimately geologists and geophysicists developed conclusive evidence for an old earth. The antiquity of the earth thus may be considered evidence in favor of the theory of evolution.

## Conclusion

Science involves a continuing interplay between hypothesis and experiment or observation. When we propose a hypothesis, we must first deduce its consequences and then seek evidence in its favor. If the hypothesis is sound and the evidence is firm, the hypothesis will likely be accepted, as in the case of continental drift. If, on the other hand, the hypothesis is faulty and the evidence is not firm, then the hypothesis will be rejected, as was the case with polywater.

In the next chapter, we will examine how pseudoscience works and show that it does not test hypotheses in a scientific manner.

### Thought Questions

1. The Roman philosopher Lucretius thought he could deduce that living organisms are composed of inanimate matter. Is such a deduction defensible? Is Lucretius's belief equivalent to a religious belief? Explain your reasoning.
2. Think of a pet "theory" that you or someone else holds. Maybe you think that it is good luck to carry a rabbit's foot or that eating fish will help you on an exam. How would you test your hypothesis in such a way as to exclude your own bias in favor of the hypothesis? Remember to state what you are looking for in advance, not after the experiment is conducted.
3. Scientists reason, in part, that the past is a guide to the future—for example, that the sun will rise tomorrow because it has done so reliably in the past. How can you justify this principle?

# Chapter 5

## *How Pseudoscience Works*

There are many more wrong answers than right ones, and they are easier to find.

—Michael Friedlander, physicist, Washington University

How can we distinguish pseudoscience from science? How can we tell that a field is pseudoscience rather than a daring and far-out yet valid approach to a specific scientific problem? The history of science is littered with heresies that were eventually accepted as correct. How do we recognize a pseudoscience and distinguish it from an honest but unlikely scientific theory?

The short answer is that the boundary between what is science and what is not science is fuzzy, and sometimes cannot be located. Indeed, philosophers of science spend a great deal of effort on what they call the *demarcation problem*, that is, the problem of distinguishing between science and nonscience. Some fields clearly do not qualify as science because their practitioners do not formulate or test hypotheses properly. But if we attempt a rigorous definition of science, we may end up considering history a science, because, for example, historians also make and test hypotheses. On the other hand, if we rule out history, then we may also have to rule out palaeontology, because it is a historical science. So let us address the issue not in terms of specific fields of science, but instead try to distinguish between science and pseudoscience.

### Properties of Pseudoscience

Science builds on its successes. We can't do chemistry, for example, without accepting the atomic theory of matter, or medicine without accepting the germ theory of disease. Pseudoscientists, by contrast, often deny accepted scientific facts or propose far-fetched hypotheses to account for unsupportable claims. Their hypotheses are not testable because they have no measurable consequences. They nevertheless defend their hypotheses at all costs and sometimes think that honest scientists are conspiring against them.

Competent scientists propose a hypothesis and try to *find out whether* it is true; pseudoscientists try to *show that* it is true. A competent scientist deduces the consequences of a hypothesis and makes a prediction that can be tested. That means a real test, with the possibility of failure. If Boris Deryagin had been a pseudoscientist, he would never have conceded that polywater was an error and would still be looking for a way to synthesize it in large quantities.

Before studying scientific creationism and intelligent-design creationism, let us examine two classic examples of how pseudoscientists operate and contrast them with Deryagin and Pauli.

### Homeopathic Medicine

*Homeopathic medicine*, or homeopathy, was developed by the German physician Samuel Hahnemann in the early nineteenth century in reaction to harsh and ineffective methods, such as blistering the skin and purging (inducing diarrhea), that were used by physicians of his time. Homeopaths apply a *law of similars*, which states that "like cures like." According to the law of similars, certain substances can cure symptoms precisely like those they cause. The reasoning is a proof by analogy: Quinine causes headaches, thirst, and fever; these are the symptoms of malaria; quinine cures malaria. As it happens, quinine truly cures malaria, but homeopaths extend the reasoning to substances that are ineffective.

Worse, Hahnemann sometimes used very toxic substances, which he diluted until they were not toxic. That is sound practice, and we do so today with substances such as digitalis and warfarin, both highly toxic chemicals (digitalis has been used on poisoned darts, and warfarin is a rat poison). Hahnemann, however, did not just dilute his solutions; rather, he thought he saw a direct correspondence between increased effectiveness and increased dilution, or decreased concentration. Indeed, homeopaths dilute their solutions so much that almost certainly not a single molecule of the medicine remains in the solution, which nevertheless contains many parts per million of impurities. They know that matter is not a continuous entity that can be diluted forever. That is, they recognize that their solutions contain no *trace* of the medicine. Instead, they claim that shaking the solution each time it is diluted *potentizes* it and leaves in the water some of the *properties* of the medicine.

Homeopathy is still practiced by medical doctors and osteopaths. What do its practitioners say about the dilution problem? They say, in effect, that the water remembers what was put into it. Beverly Rubik, a physician who promotes so-called alternative medicine at the Institute for Frontier Science in Oakland, California, says, "We need to rethink the properties of matter. It's that deep." Rubik, who did her work on homeopathy at Johns Hopkins University, proposes no new theory of matter and does not understand that a new theory has to be consistent with successful, existing theories. That is, it is extremely difficult to overthrow a theory that is known to work; usually, the best we can do is to extend such a theory. Homeopathy, as understood by Rubik, does not extend modern chemical knowledge; to the contrary, it denies our present understanding of the chemical bond.

Do we have any evidence that the water remembers? Jacques Benveniste, a researcher at the University of Paris, thinks we do, and he undertook a series of experiments to attempt to show that it does. Specifically, Benveniste exposed white blood cells to highly diluted homeopathic solutions of a certain antibody. If the cells responded to the solutions, they would have absorbed a certain red dye differently from cells that did not respond. Benveniste and his technician observed the cells in a microscope and thought they saw a response. Their result was reported in the authoritative science journal *Nature* and was met with widespread disbelief.

In response to criticism, *Nature* sent a team of outside investigators to Benveniste's laboratory. These investigators concluded that Benveniste's technician knew in advance what he was expected to find—and found it. Benveniste did not watch out for *experimenter bias* (or *confirmation bias*): The observer or experimenter knows what to look for and seems to find it, whereas an observer who is untainted by such knowledge does not. When the team arranged a *blind test*, that is, a test in which Benveniste's technician did not know what to look for, the results were negative. Benveniste's results look like a classic case of self-delusion.

Benveniste hypothesizes that molecules communicate by electromagnetic radiation instead of by exchanging electrons. The radiation remains in the water, even after the molecules are removed by dilution. His theory makes no predictions that can be tested. It does not explain why, if the radiation stays in the water, it is not diluted as the solution is diluted, nor does it postulate any mechanism that might prevent the radiation from escaping from the solution; electromagnetic radiation normally travels at the speed of light.

Benveniste and Rubik, despite their advanced degrees, are practicing pseudoscience, Rubik by pretending that we do not understand chemical bonds, and Benveniste by inventing an auxiliary hypothesis designed to explain away the dilution problem. They do not seem to understand that patients sometimes appear to respond to a worthless drug or even a sugar pill (a *placebo*), just because the patients believe it works. It is very easy for either a patient or a practitioner to be deceived by a "cure" effected by a disease's simply having run its course. Indeed, that is why drugs are subjected to rigorous testing before they are approved for the market.

### Astrology

Astrologers think that the positions of the heavenly bodies among the stars have a direct influence on human behavior or personality. Some astrologers attempt predictions on the basis of astrological signs. Physical scientists reject astrology because they can see no mechanism that might cause the positions of the heavenly bodies to directly affect the behavior of humans on earth. The nearest such bodies, the sun and the moon, can cause oceanic tides, but tides are unimpressive in the sense that the oceans are several kilometers deep, and the tides rise and fall only a few tens of meters at most. Further, tides affect only deep bodies of water, not shallow, because the tide depends on the difference between the gravitational attraction at the surface and at the bottom. The difference of the moon's attraction between the top of your head and your feet is immeasurable, even if the moon is directly over your head, so it is very unlikely that a tidal force could affect you in any way. Stars or planets influence even the tides in the ocean immeasurably.

Additionally, astrologers use obsolete data when they construct their charts. Owing to the slow rotation of the earth's axis, the alignment of the sun, moon, and planets among the constellations gradually drifts with respect to the seasons. When astrologers construct your birth chart, they use the alignment of the planets *as they appeared on your birthday 2,000 years ago*. If astrology had any physical basis whatsoever, astrologers would have to use the alignments of the planets today, not at the time of Julius Caesar.

Geoffrey Dean, a prominent, self-described "serious" astrologer, claims that astrology has been empirically demonstrated to work. Dean does not consider newspaper astrology columns to be serious astrology, but he claims that there are about as many serious astrologers as there are psychologists in the United States. According to Dean, a majority of Western astrologers concentrate on psychology and counseling, not prediction.

How well does astrology perform as a counseling tool? Dean gives the example of an astrologer who discusses a meek person who has five planets in Aries. People with all five planets in Aries ought to be very aggressive, according to Dean. Thus, the astrologer looks for other signs that indicate meekness. That is, he looks for an auxiliary hypothesis with which to explain away his failed prediction. If he cannot find such signs, he employs another auxiliary hypothesis, specifically, the hypothesis that sometimes the opposite of what he predicts also happens: "if a person has an *excess* of planets in a particular sign, he will tend to suppress the characteristics of that sign because he is scared that, if he reveals them, he will carry them to excess. But if on the next day I meet a very aggressive person who has five planets in Aries, I will change my tune: I will say that he *had* to be like that because of his planets in Aries [italics in original]." In other words, if someone has five planets in Aries, then, as a direct result, he will be either aggressive or not aggressive. Since everyone is either aggressive or not aggressive, or sometimes aggressive and sometimes not aggressive, the prediction is necessarily accurate, but it also is not testable and is therefore valueless. If we know that someone has five planets in Aries, then we know absolutely nothing meaningful about that person.

Dean's astrologer attributes the failure of his prediction to *reaction formation*, a term we borrow from another pseudoscience, Freudian psychoanalysis. That is, the person with five planets in Aries formed a reaction and reacted oppositely to what the astrologer expected. A theory that can explain everything in terms of reaction formation explains nothing, because we can always say that our prediction failed because of reaction formation. A theory that cannot be refuted is not testable. It is not scientific. It is pseudoscience.

Indeed, Dean describes a number of tests, such as giving two astrological charts to each subject and asking the subjects to pick which is theirs. Dean cites no study in which subjects could pick out their own charts more often than would have occurred by chance. Subjects can recognize themselves in valid personality tests, however, so the problem is with the astrology, not with the subjects or the methodology. Dean has partly recanted and now says that astrology is "untrue." Incredibly, however, he still concludes that astrology can help people as long as it "satisfies" them. But he provides no evidence that astrology is effective even in this limited way.

### Conclusion

Competent scientists propose testable hypotheses to explain observations. They test their hypotheses repeatedly in the court of nature. If a hypothesis fails, they will either abandon it or develop an auxiliary hypothesis, which is itself testable. They do not stick stubbornly to failed or outmoded theories.

Pseudoscientists, by contrast, "know" the answer before they propose the hypothesis. They do not try to find out whether a proposition is true but rather assume the answer and so are vulnerable to experimenter bias. They prop up their hypotheses by vague references to unknown radiation or by denying a well-known scientific fact. If these devices do not work, then they propose an auxiliary hypothesis such as reaction formation, which makes their claims untestable and therefore unprovable.

In subsequent chapters, we will examine creationism, especially intelligent-design creationism, and decide whether it is science or pseudoscience.

## Thought Questions

1. Think of an example of a "testable hypothesis." Write it down along with several ideas for how to test your hypothesis. Compare your methods with those of other students.
2. Think of an example of an "untestable hypothesis." Can you devise honest ways to test it? Could it be tested if, for example, our technology improved?

## Chapter 6

# *Why Creationism Fails*

> My answer to him was, "John, when people thought the earth was flat, they were wrong. When people thought the earth was spherical, they were wrong. But if you think that thinking the earth is spherical is just as wrong as thinking the earth is flat, then your view is wronger than both of them put together."
>
> The basic trouble, you see, is that people think that "right" and "wrong" are absolute; that everything that isn't perfectly and completely right is totally and equally wrong.
>
> —Isaac Asimov, writer and polymath

This chapter discusses scientific creationism, under which heading we classify both young- and old-earth creationism. Young-earth creationists assume that the earth and indeed the universe are only thousands of years old, whereas old-earth creationists agree that the earth is billions of years old. Both use the Bible as their source of information, and both develop supposedly scientific arguments to reconcile what we know of science with their interpretations of the Bible and, in particular, the first few chapters of Genesis.

Young-earth creationists, for example, search high and low for evidence of the Noachian flood and invent fake fields of study such as flood geology. Flood geology claims that the variously aged fossil remains we find in the earth were laid down roughly simultaneously, when plants and animals died in Noah's flood.

Fossils are remains of plants or animals in which the tissues have been replaced by minerals. With time, fossils are covered by sediment, so older fossils are buried deeper than more recent fossils. That is, the fossil record is layered, or *stratified*, precisely because some fossils were deposited many thousands or millions of years before others. Radiocarbon dating and other forms of *radiometric dating* (see chapter 15) confirm that all fossils were not deposited within a single, short time.

If flood geologists are right, then we would not expect fossils to be deposited in layers. To explain the layering that is actually observed, flood geologists point to some unspecified hydrodynamic forces and claim that these forces deposited animals sequentially according to their size, shape, or buoyancy. Others claim that more-advanced animals were better able to escape the flood; hence their bodies were deposited above the others'. It is hard to see, however, how a more-advanced chimpanzee, deposited on a higher layer, was more able to escape a flood than a less-advanced crocodile or fish, and especially hard to see how a more-advanced maple tree was more able than a less-advanced conifer or fern to escape. Yet the fossils of conifers and ferns, which evolved long before broadleaved trees such as maples, are found much deeper than any broadleaved trees. The flood geologists propose no testable hypotheses that might distinguish between flood geology and conventional geology; their unspecified hydrodynamic forces are no more than an auxiliary hypothesis designed to prop up their failed theory.

Fossil fuels, such as coal and oil, also count against the flood geology: It took millions of years to deposit enough organic material to account for all the fossil fuels found under the surface of the earth. If all known deposits of fossil fuel had been deposited in the year or so following the flood, then the earth must have been very fruitful indeed, for the earth does not grow enough organic matter today to account for even a tiny fraction of the known fossil fuel reserves.

Finally, it is hard to see where all the water came from. The authors of the Book of Genesis thought that the sky was a dome that covered the (flat) earth and that water occupied the region above the dome. So they had no intellectual problem believing that the dome was opened and the water allowed to spill onto the earth. We, on the other hand, know that the appearance of a solid dome is an illusion brought about by the scattering of sunlight by the atmosphere. We therefore have to ask where the water came from. There is no satisfactory answer.

Young-earth creationism thus fails because its claim that the earth is a few thousand years old is, frankly, contrary to fact. In place of evidence, young-earth creationists supply vague hydrodynamic forces that cannot satisfactorily account for the source of the water for a worldwide flood, the stratification in the fossil record, or the existence of fossil fuels.

### Old-Earth Creationism

Old-earth creationists generally accept the fossil record and the antiquity of the earth. They believe literally in the creation myth of Genesis, however, and find ways to force it into apparent agreement with modern science. There are subtheories within old-earth theory. Day-age theorists argue that a day means an age; gap theorists argue that life was created at once on a preexisting earth after an unspecified gap following the creation of the earth. Many of the beliefs of old-earth creationists are scientifically neutral, and we will not dwell on them or their beliefs. Other old-earth creationists, however, do immeasurable harm by twisting scientific evidence to fit their preconceived ideas or by portraying evolutionary biology or geology as a collection of half-baked ideas that are merely stated and have no support.

The physicist Gerald Schroeder, for example, argues that the first chapter of the Book of Genesis was written by a being who lives in what Schroeder calls *logarithmic time*. In logarithmic time, the first "day" after the creation is 8 billion years long; the second is 4 billion years; the third is 2 billion years; and so on. Only at the end of the first chapter of Genesis is the writing turned over to humans who live in linear, or nonlogarithmic, time. Schroeder claims that the chronology of Genesis makes sense in logarithmic time.

The superficial agreement between scientific fact and the six days of creation is fairly good, but there are problems. In the Book of Genesis, water is above the sky (remember the flood!), dry land and plants are created before the sun, and fowl are created before reptiles. Schroeder notes that flowering plants would not have appeared on the third day but on the fifth, but solves his dilemma with an auxiliary hypothesis (see chapter 4) borrowed from the medieval Jewish sage Nachmanides: Flowering plants *developed* during the next days. What did they develop from? Single-celled plants. Sounds a good bit like evolution, but Schroeder presents no evidence to support the claim that single-celled plants were actually created on the third day.

Schroeder disagrees with the flood geologists and admits that there is no geological evidence for the Noachican flood, so he suggests that the flood was only local. Here he disputes the clear word of the Bible itself: "And the waters prevailed exceedingly upon the earth; *and all the high hills, that were under the whole heaven, were covered*. Fifteen cubits upward did the waters prevail; and the mountains were covered" (Genesis 7:19–20, King James Version, emphasis added). He uses a far-fetched linguistic argument to justify his claim that the biblical authors intended the flood to be local, but he provides no good support for that argument. Additionally, at the beginning of Schroeder's sixth day, the fossil record shows a major mass extinction, but it is not mentioned in Genesis, and Schroeder all but ignores it. Schroeder accepts evolution on the whole but argues that God gave humans souls at the time of Adam and Eve. Thus, Adam and Eve were the first truly human beings; only after humans were endowed with souls did they develop writing and civilization. Schroeder calls the other people who were around at the time *nonhuman hominids*. Unhappily, the fossil record cannot distinguish between human beings and nonhuman hominids, so Schroeder's hypothesis is untestable.

The astrophysicist Hugh Ross, another old-earth creationist, similarly accepts uncritically the Bible's claim that human lifetimes were typically 900 years around the time of Adam. Indeed, people were forbidden to eat meat because toxins contained in the meat, but not in plants, would have poisoned

The Bible as a Science Book

It is very dangerous to use the Bible to "prove" or "verify" modern science: It's a game that two can play. For example, 1 Kings 7:23 says, "And he made a molten sea, ten cubits from the one brim to the other: it was round all about, . . . and a line of thirty cubits did compass it round about." According to the logic of the biblical literalist, pi must have been precisely equal to 3 in the days of Solomon.

Similarly, the Hebrew Bible has about a dozen references concerning the heaven above and the earth below, the ends of the earth, the corners of the earth, and the length of the earth. Again, according to the logic of the biblical literalist, the earth must have been flat and rectangular in biblical times.

The earth was not flat, and pi was not precisely equal to 3. The earth looked flat locally, and pi was approximately equal to 3; the authors of the Bible were recording what they saw and knew several thousand years ago, at the dawn of civilization. They were not writing science or mathematics or geography, and the Bible cannot be used to settle questions in those fields. See the section on overlapping and nonoverlapping magisteria in chapter 18.

them well before their allotted 900 years were up. Evidently God was unaware at the time of the cancer-causing chemicals in mushrooms. Later, because people had become wicked and could do a lot of damage in 900 years, God used cosmic rays from the Vela supernova to shorten their lives. Unless the fossil record can distinguish between healthy 900-year-olds and healthy 90-year-olds, Ross's hypothesis is likewise untestable.

Old-earth creationism fails not because it makes scientific claims that are demonstrably contrary to fact but rather because it feels compelled to force-fit the chronology of the Bible to the known chronologies of geology and evolution. Not only is the fit not good; the evidence in its favor is also lacking.

## Conclusion

Scientific creationism has all the earmarks of a pseudoscience. Its proponents know in advance the answer they want, and they formulate untestable hypotheses to obtain those answers. Thus, young-earth creationists develop a theory of the Noachian flood that purports to account for the stratification in the fossil record. Their theory provides no testable hypotheses but rather refers to unspecified hydrodynamic forces and vague references to more-advanced and less-advanced animals. Old-earth creationists are less harmful to science because they do not deny cosmology and the fossil record, but they nevertheless propose untestable hypotheses designed to prop up their prior beliefs. They are not practicing science but rather rationalizing a religious belief.

In the next chapter, we will look at a more respectable philosophical argument, the argument from design, which argues by analogy that the universe or some of its components appear to have been designed, and that therefore they were designed. Though it is most likely wrong and can be misused, the argument from design is not pseudoscience and indeed was the most plausible way of accounting for many observed facts of biology before the development of the theory of evolution.

### Thought Questions

1. Consider the quotation with which we began this chapter. Can you think of any statements that are nearly right (or slightly wrong) but still useful? What analogous statements would be very wrong or wholly wrong?
2. Compare the chronology in the first chapter of Genesis with what we know from cosmology and paleontology. Compare the chronology in the first chapter of Genesis with the chronology in the second chapter. Do you note any discrepancies or inconsistencies? How can you account for these?
3. Name some professions that make and test hypotheses. Are they necessarily scientific? What makes one profession scientific and another not?
4. Albert Einstein rejected a common interpretation of quantum mechanics because he could not accept that the universe worked that way, that is, could not be random at the deepest levels. This approach is sometimes called the *argument from incredulity*. Do creationists who reject evolution necessarily use the argument from incredulity? Do atheists who reject a deity? Are those who use the argument from incredulity necessarily wrong? Explain your answers.

## Chapter 7

# *The Argument from Design*

> In crossing a heath, suppose I pitched my foot against a *stone*, and were asked how the stone came to be there; I might possibly answer, that, for any thing I knew to the contrary, it had lain there for ever: nor would it perhaps be very easy to show the absurdity of this answer. But suppose I had found a *watch* upon the ground, and it should be inquired how the watch happened to be in that place; I should hardly think of the answer which I had before given, that, for any thing I knew, the watch might have always been there. Yet why should not this answer serve for the watch as well as for the stone? Why is it not as admissible in the second case, as in the first? [Because], when we come to inspect the watch, we perceive (what we could not discover in the stone) that its several parts are framed and put together for a purpose, *e.g.* that they are so formed and adjusted as to produce motion, and that motion so regulated as to point out the hour of the day; that, if the different parts had been differently shaped from what they are, of a different size from what they are, or placed after any other manner, or in any other order, than that in which they are placed, either no motion at all would have been carried on in the machine, or none which would have answered the use that is now served by it.
>
> —William Paley, clergyman and naturalist

These words by William Paley were written in 1800. They are the classic statement of the argument from design.

In Paley's time, most people believed in fixed species; that is, they thought that the species were set at the creation and remained unchanged thereafter. Many scientific creationists hold that view today, arguing that *kinds* (see chapter 3) were created at once, and that, while there can be variation within kinds, there can be no crossing from one kind to another. Thus, the dog kind may originally have had only one or two members but now displays a number of breeds. In short, some scientific creationists accept *microevolution*, or slight changes within kinds, but reject *macroevolution*, or evolution from one species to another. Others may admit that wolves, foxes, and dogs, for example, have descended from the canine kind but would deny that the reptile kind has descended from the amphibian kind.

Theories of evolution before Darwin were rudimentary at best; Darwin's grandfather, Erasmus Darwin, proposed that animals descended from a single ancestral species and continually improved themselves in some unspecified way as a result of their own activities and experiences. After Paley published his book, Jean-Baptiste de Lamarck in 1809 developed a theory of *inheritance of acquired characteristics* and postulated that organisms improved themselves by passing to their descendants characteristics that they acquired during their lifetimes. Lamarck assumed that only characteristics that the animal acquired as a result of willing them were passed on, so, for example, bulldogs were not born with their tails cut off, even though their ancestors' tails had been cut off for generations. Lamarck's theory is completely consistent with the then-accepted view that some kind of *life force* propelled living organisms and was not as unreasonable at the time as it is sometimes portrayed today.

Thus, until 1858, when Darwin and Alfred Russel Wallace published a comprehensive theory of natural selection, it would have been hard not to accept Paley's argument, because Paley was an excellent naturalist, and his argument was very carefully thought out. Darwin and Wallace, however, made the need for a designer wholly unnecessary by providing a naturalistic theory that accounted for the observed facts without a designer and accounted for some of those facts better than a designer.

Paley's argument, which was good in its time, has all the concepts we will see in modern creationism: The watch has a purpose, it is intricately designed, all the parts are well matched, and if one part is out of place, the watch ceases to function. Paley assumes that the universe has a purpose; maybe it does, maybe it does not. If it does not, then Paley's argument fails.

His argument fails on other accounts as well. For instance, a watch is obviously artificial. It is made by a human being (or several human beings). We know what devices made by human beings look like, so we can deduce the artificiality of the watch. What does an artifact made by a supremely powerful being like God look like? No one really knows; hence we cannot fairly infer a designer for a rock, say, in the way we can with a watch.

To infer design, we have to know something about the designer. If we look at a picture of Mount Rushmore, we will immediately recognize it as designed. Why? Because, for one thing, we recognize intricate human faces when we see them. A space alien who had never seen a human face would not have the slightest idea that Mount Rushmore was designed. Even if the alien got close enough to see the toolmarks, he could not meaningfully infer design. All the alien would know was that some unknown agent, possibly natural, had carved the rocks in a repeating pattern. A pigeon might think that Mount Rushmore had been cut away to make nesting places for pigeons.

We do not think that an ancient hunter-gatherer, if she found a watch, would necessarily infer that it was a human invention. To the contrary, she would not have the vaguest idea what it was, precisely because she could not infer a purpose. She would have to be familiar with sundials, at least, before she could deduce the function of a watch; we simply cannot infer a human designer unless we know what an artifact is used for. Indeed, we recognize artifacts such as stone tools only because we know ahead of time what they are; an amateur may easily mistake a randomly damaged stone for a stone tool or an arrowhead.

Still, if we examine the watch closely, we will see tool marks. We may be able to figure out how the watch works and deduce its purpose, provided that we belong to an advanced enough civilization. The watch may even have an inscription such as "Made in Switzerland." We can go to Switzerland, find a watchmaker, and verify the existence of watchmakers. None of these is true of the earth or the universe: We have never identified any tool marks, we do not know the purpose and have not deduced any purpose, and no one has ever found a 4.5-billion-year-old stone artifact (in the right geological stratum) with the words, "Made by God."

We consider design to be a more important criterion than purpose. Something can be designed without purpose. Thus, if we happened upon a sand castle, we would deduce that it had a designer, even though it possibly had no purpose, or at least no lasting purpose, other than the amusement of the designer. By tacking purpose to design, Paley has made his task more difficult. After all, it is possible that God designed the universe with no lasting purpose.

## Conclusion

When William Paley proposed his version of the argument from design in 1800, there was no compelling theory of evolution, and most people thought that species had been fixed at the time of the creation. Paley's argument from design supported people's preconception and made sense in its time. Still, evolution was in the air around 1800, even though no one had thought of a convincing mechanism. The argument from design was no longer needed after Darwin and Wallace developed the theory of natural selection, because that theory accounted for the observed facts without recourse to divine intervention.

In the next chapter, we will discuss the current state of the art in creationism: intelligent-design creationism. Intelligent-design creationism is the modern successor to the scientific creationism of chapter 6, but it is also in a way the successor to the argument from design. It is based on a pseudo-mathematical argument by the mathematician and theologian William Dembski

### Infinite Regression

The origin of this modern parable is unknown, but it always involves turtles and has been attributed to great scientists or philosophers ranging from Thomas H. Huxley to Bertrand Russell. The parable goes like this:

A scientist is giving a lecture on the origin of the solar system, when a member of the audience, typically described as a little old lady, informs him that the earth is a flat plate that rests on the back of a giant turtle. When the scientist asks on what the turtle stands, the person replies that it stands on the back of another, larger turtle. That turtle stands on the back of a yet larger turtle. When questioned further, the person replies, "It's turtles all the way down."

Turtles all the way down has become a metaphor for what is known as an infinite regression (or regress). If indeed each turtle stood on the back of another turtle with no known solid ground or platform on which to rest, then there would have to be an infinite number of turtles. Each turtle depends on an infinite number of turtles below it. If each turtle has to be bigger than those above it, then the hypothetical bottommost turtle must be infinitely large.

The argument from design suffers from the turtle problem: If there was a designer, then who made that designer? Another designer, even more powerful than the first? If so, then who made that designer? We are left with an infinite regression with an infinite number of designers, the hypothetical last of which must be infinitely powerful.

An infinite regression seems absurd, but it is still possible that the universe has existed forever (see chapter 15). If, however, we reject the possibility of an infinite age, then we must admit that something can be created out of nothing. That something must be either the universe itself or its designer. A philosophical principle called the *principle of parsimony*, or *Ockham's razor*, says that we should adopt the simplest workable solution, in this case, that the universe was created out of nothing. That is the simpler solution because the early universe was formless and void, whereas a designer would have had to be extremely complex.

Even if we assume that the designer, rather than the universe, was created out of nothing, we might infer that that designer was not supernatural, as some intelligent-design creationists coyly suggest. They have a point. Ockham's razor directs that, if we have to choose between a natural and a supernatural designer, the more parsimonious solution is that the designer was not supernatural. Indeed, if the designer is not supernatural, then we may hypothesize that the designer evolved by natural selection and possibly can be studied scientifically.

and on the idea that some biological structures are so complex that they cannot have evolved gradually.

Thought Questions

1. Name some items that are designed. How do you know that they are designed?
2. Distinguish between design and purpose. Do all the items you named in question 1 have a purpose? How does purpose help in assessing whether an object has been designed?
3. If you found a mohaskinator that had been designed and built by a man from Arcturus, how would you know it was designed? That is, what would you look for to infer a designer?

CHAPTER 8

# Why Intelligent-Design Creationism Fails

This isn't really, and never has been, a debate about science. It's about religion and philosophy.

—Phillip Johnson, University of California Law School

INTELLIGENT-DESIGN CREATIONISM is the successor to scientific creationism such as that practiced by the flood geologists. ID creationists argue, in a variation on the argument from design, that they can rigorously infer design, whereas William Paley more or less assumed design. The heart of their argument is the claim that organisms are too complex to have evolved by chance. ID creationism has two main planks: a logical procedure called the *explanatory filter*, and analogies between living organisms and machines. The explanatory filter is said to never wrongly infer design where there is none. The analogy with machines is used to argue that cells have so many complex, interacting parts that they could not have been assembled, except by design. The two arguments are closely related and rely on intuitive notions about probability.

## EXPLANATORY FILTER

The explanatory filter is a logical procedure that claims to use probability to infer whether an artifact is natural or designed. It is the brainchild of the mathematician William Dembski. Dembski claims that the filter can rigorously infer design and further, though it might sometimes miss an instance of design (a *false negative*), that it will never demonstrate design where there is none (a *false positive*). False negative and false positive are statistical terms: We get a false positive when we perform a test and incorrectly infer a positive result, as when we seem to detect tuberculosis in a person who in fact does not have tuberculosis. Similarly, we will get a false negative if we fail to detect tuberculosis in a person who has the disease.

Dembski's filter is shown in figure 2. To illustrate, let us apply it to some artifact, such as a snowflake or a mouse. Dembski calls the snowflake or the

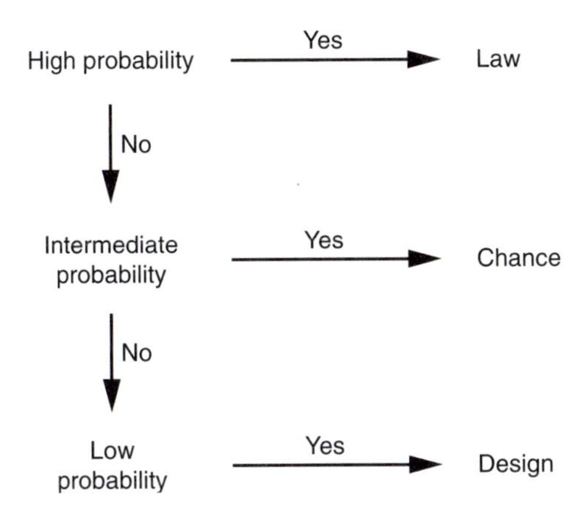


2. Dembski's explanatory filter. The probabilities are not given, so the filter cannot be used in practice.

mouse an *event*. If an event happens with regularity and very high probability, we attribute the event to a law and conclude that it is not designed. If the event happens with intermediate probability, then we attribute its occurrence to chance. If, however, the event happens with very low probability, less than $10^{-150}$, and if it corresponds to a *meaningful message* (see below), then we attribute the event to design.

The explanatory filter is, unfortunately, unusable. How high is high probability? Dembski doesn't say. How high is intermediate? Again, he doesn't say. How, then, do we navigate the filter? We cannot get past the first branch point if we do not have a quantitative number for high probability.

Additionally, Dembski provides no convincing proof that a probability of $10^{-150}$ necessarily infers design. It could, rather, infer "don't know" or even "haven't the foggiest." Indeed, deciding that an event is designed may automatically prevent us from seeking and therefore finding the real cause. "Don't know" is by far the better conclusion, because it is a spur to action: Find out.

Finally, the filter is by no means immune to false positives. Here is a real-world example in which the police, in effect, applied the explanatory filter with tragic results. A British woman lost three babies to sudden infant death syndrome (SIDS, or crib death) within four years. The Crown Prosecution Service reasoned as follows: One death is tragic; two deaths are suspicious; three deaths are murder. The woman was prosecuted.

What the prosecutors did not know or ignored was that SIDS may be a genetic disease that runs in families. Indeed, the woman's grandmother testified that three of her own children had died of unexplained causes before the ages of six weeks (in the 1940s, before SIDS had been recognized). A geneticist testified that SIDS could indeed run in families and suggested two possible mechanisms. The jury used this background information—what Dembski calls *side information*—and acquitted the woman.

In short, the prosecutors applied the explanatory filter and very probably got a false positive, whereas the jury looked at the side information and drew the opposite conclusion. Dembski argues that the side information must be ignored if the filter is to be used properly. The forensic archeologist Gary Hurd argues to the contrary that only the side information is relevant. An archaeologist examining some shells to see whether they were used as beads does not apply the explanatory filter. Rather, she looks at the patterns of wear to see whether they were drilled or caused by natural mechanisms, she examines the holes in the shells to see whether they could have been caused by predators, she looks for signs of paint, and so on. If, for example, the patterns of scratches in the shells are regular, if the shells are all the same size, if the holes are all in the same places on the shells, if traces of paint are found, then the archaeologist infers that the shells are beads. She does not use probability or the explanatory filter but rather formulates hypotheses based on the side information and then decides whether the evidence points to natural or manufactured origin.

We can't prove that the prosecutors were wrong when they charged the woman with murder. This real-world example does, however, show that the explanatory filter is by no means immune to false positives—precisely because we can never know when we have all the side information, and this information is often crucial, not *side* information at all. Dembski claims that his filter is immune to false positives. But if the explanatory filter cannot be applied to a simple case such as identifying beads or ferreting out a murderer, it is equally useless for identifying the artifacts of a designer whose habits and intentions are wholly unknown to us.

### SIDE INFORMATION

To illustrate the critical importance of side information, consider the following experiment: We set up a cannon, aim it at a target, and fix it permanently in position. We fire the cannon a number of times, always using the same charge, and mark where the cannonballs land. We do not expect to hit the target exactly because of uncertainty in the velocity and direction of the cannonballs as they leave the barrel of the cannon. The uncertainty is caused primarily by the fact that the cannonball is slightly smaller than the bore of the cannon; variations in the mass of the cannonball and the size of the charge are negligible, as is wind velocity. Thus, we expect to see a more or less uniform distribution of marks around the target, as in figure 3, left. Instead, we get the distribution shown in figure 3, right, where the cannonballs are not located randomly about the target but rather lie in a long, narrow region. What could have caused such a distribution? The answer is given in figure 6, at the end of this chapter. Don't peek without trying to figure it out first!

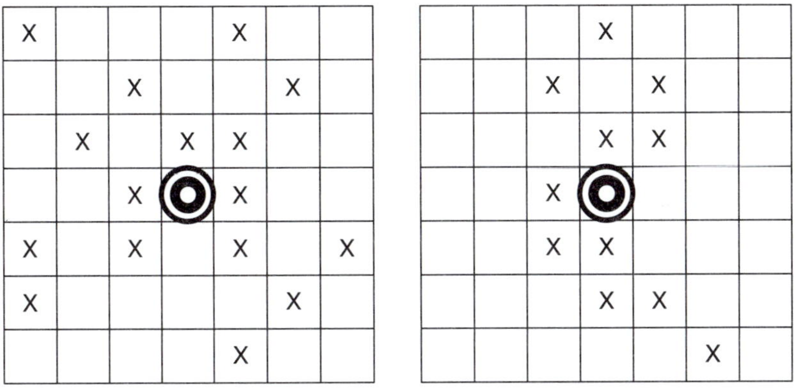

3. Left: Expected distribution of cannonballs. Right: Measured distribution. Why is the measured distribution of cannonballs oriented along the vertical axis?

### Dembski's Arrow

To illustrate the supposed relation between low probability and design, Dembski uses an analogy. Consider an archer, says Dembski. If the archer aims at a target on a wall and hits the target repeatedly, then we may conclude that the archer was highly skilled and did not simply shoot arrows randomly at the wall. To Dembski, the skill of the archer is analogous to design: If the archer hit the target over and over, he did so by design and not by accident. Thus, by looking at the pattern of repeated hits, we can infer design as opposed to chance. By contrast, we would infer chance if we saw arrows scattered all over the wall and only one or a few stuck into the target.

Dembski intends his archer to be an analogy for the evolution of different complex biochemical systems such as the chlorophyll molecule, which he claims is so unlikely (remember the filter) that it must have been designed. There are two main problems with Dembski's analogy.

First, no one ever suggested that chlorophyll or anything else evolved entirely by chance. To the contrary, chlorophyll evolved by a combination of law and chance. Indeed, Dembski understands full well that complicated artifacts can be produced by law and chance. Take one of his favorite examples: the snowflake. The snowflake crystallizes as it falls because of physical laws, but its exact form depends on random variables such as the temperature and humidity of the air through which the snowflake has fallen. Snowflakes form by physical laws, but each snowflake is different from each other because of chance. The probability of any one pattern is very small. Dembski correctly does not infer design in a snowflake, even though the pattern of any given snowflake is very improbable. He fails, however, to apply the same logic to the evolution of chlorophyll.

### More on Side Information

William Dembski defines a pattern as some recognizable sequence, such as a series of digits or other symbols. A pattern contains *complex specified information* and cannot be the result of natural causes if the probability of that pattern's appearance is less than a certain value. Dembski stresses that we have to examine the pattern in isolation and free of any background information, or side information. To illustrate the importance of side information, let us consider one of Dembski's favorite examples, a county clerk who put Democrats on the top line of the ballot 40 out of 41 times. We represent his action as a sequence:

DDDDDDDDDDDDDDDDDDDDDDRDDDDDDDDD
DDDDDDDDD

where D stands for Democrat and R stands for Republican.

According to Dembski, we have to look at the sequence in isolation and decide whether or not it is too improbable to have occurred by chance. The information about the clerk is side information and may not be considered; that is, it must be *detached* from the pattern. Thus, we look at the sequence and presume that the odds on each occasion are 50–50 that the clerk will choose a Democrat if he is choosing fairly. The chance of choosing a Democrat 40 out of 41 times, regardless of the order, is about 1 in 50 billion. Thus, it seems likely that the clerk cheated, and indeed he was chastised by a court.

Is it really true, though, that the sequence is too improbable to have been generated by natural causes? Following mathematicians Wesley Elsberry and Jeffrey Shallit, let us ask, "What if the sequence represents not an election but rather dry and rainy days over a 41-day period in Boulder, Colorado?" Dry for 22 days, rained 1 day, dry for another 18. Nothing unusual about that. Surely we can imagine many other ways whereby the sequence could have been generated by natural causes. Dembski mistakenly thinks that the sequence is exceedingly improbable because, ignoring his own standards, he has neglected to detach the side information from the sequence. Without the side information about the clerk, we would not think the sequence is necessarily improbable. As Gary Hurd has noted in connection with the explanatory filter, the side information is critical. It cannot be detached from the sequence.

Dembski has proven only that you will probably not get 40 heads in 41 tosses of a fair coin. But we knew that.

Dembski apparently thinks that evolution is directed and that chlorophyll is some kind of predetermined end point, which he calls a *specification*. Indeed, the only difference between a snowflake and a chlorophyll molecule is that the chlorophyll molecule is specified, but the snowflake is not. Thus, a specification means merely that Dembski has some special interest in the outcome.

At any rate, chlorophyll is not a predetermined end point. To the contrary, there are several kinds of chlorophyll, so we can't talk about the probability of evolving a plain old, single chlorophyll molecule. In addition, chlorophyll is not the only molecule that enables photosynthesis; a molecule called bacteriorhodopsin performs a similar function in archaebacteria. Finally, some plants in the deep-sea vents use *chemosynthesis* rather than photosynthesis for energy. Possibly other, unknown mechanisms could have evolved but did not. There is more than one way to skin a cat, and there is more than one way to power a cell.

Thus, Dembski's wall, for all we know, may be covered with targets: known and unknown types of chlorophyll, of bacteriorhodopsin, of chemosynthesis—as well as other unknown mechanisms. Probably the wall is not so densely covered that the archer cannot miss, but maybe it is densely enough covered that he will hit many targets after taking hundreds of millions of shots over hundreds of millions of years.

### Behe's Mousetrap

Old-fashioned creationists used to ask, in rebuttal of the concept of descent with modification, "What good is half an eye?!" Indeed, Darwin himself worried about the evolution of an organ as complex as an eye and noted that if anyone could show that the evolution of the eye was impossible, his theory would be cast into serious doubt.

Darwin need not worry. First, half an eye is far better than none. A woman we will call Aunt Rita had age-related macular degeneration, a condition that destroys the central vision and leaves the patient with a visual acuity of perhaps 20/200 (see also chapter 14). Yet she could walk to the store, navigate her kitchen, read an especially large telephone dial, and do countless other tasks that require only limited vision. In contrast to Aunt Rita, some people are born wholly without color-sensing cones (achromatopia) and so have no central vision whatsoever. They have an advantage over Aunt Rita, because they learn to use peripheral vision effectively and often can see better in dim light than people with normal vision. Neither has a "whole" eye in the sense that the creationists mean the term, yet their eyes are far better than none.

The eye is soft tissue and does not fossilize, so we have no fossil record of its evolution. Nevertheless, we can study animals across many taxa and get a very good idea how the eye evolved. The simplest eyes are light-sensitive eyespots that allow a primitive organism, for example, to detect the presence of

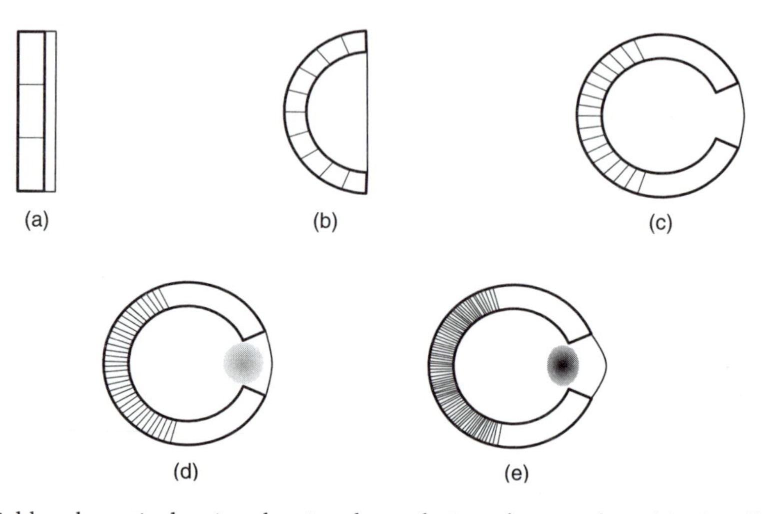


4. Highly schematic drawing showing the evolution of an eye from (a) a handful of light-sensitive cells covered by a transparent membrane through (b) a concavity that provides directional information to (c) a pinhole eye, (d) an eye with a weak lens, and finally (e) an eye with a cornea and a fully formed lens. After Nilsson, Dan-E., and Suzanne Pelger. 1994. "A pessimistic estimate of the time required for an eye to evolve." *Proceedings of the Royal Society of London B*. 256: 53–58.

a predator. Those eyes evolved into concave eyes that give directionality and therefore more-useful information than a simple eyespot (figure 4). Next, the concavity closed over, and a cornea formed. Finally, in species with a "camera eye," a lens formed in order to allow variable focusing. A sophisticated calculation by Dan-Erik Nilsson and Suzanne Pelger of Lund University in Sweden showed that a camera eye can evolve from a simple eyespot in a few hundred thousand years—the blink of an eye (pun intended) in geologic time.

If a primitive light-sensitive eyespot is a tenth of an eye, let alone half an eye, then a tenth of an eye is substantially better than none, and we know approximately how a camera eye evolved. Unwilling to give up on the design argument, the biochemist Michael Behe of Lehigh University applied it to an artifact nearer to the cutting edge of science: the bacterial flagellum. What good, he asks, is half a flagellum?

Behe argues that any system consisting of three or more well-matched parts could not have evolved without a designer because it is too unlikely that three (or four or five) parts will spontaneously come together and perform a complex function. Behe calls such a system *irreducibly complex*. Behe's irreducible complexity is closely related to Dembski's low probability, which Dembski sometimes calls *specified complexity*. For both Dembski and Behe, if a system has several critical, interacting parts, it must have been designed.

EVOLUTION OF THE MOUSETRAP

Michael Behe likes to talk of "the" mousetrap as much as he likes to talk of "the" flagellum. But there is no one mousetrap, any more than there is one flagellum. Niall Shanks, a philosopher and historian of science now at Wichita State University, notes that the spring-loaded device that we now think of as a mousetrap was patented in 1903 and borrowed ideas from a half-dozen earlier patents. Indeed, both before and after the patent was issued, mice were stabbed, beheaded, choked, electrocuted, clubbed, and drowned in various kinds of traps. Some traps trapped the mice alive, presumably in the mistaken belief that they could be released and survive outside the house. One particularly ghastly mousetrap used a modern glue to immobilize a mouse until it died of exhaustion or dehydration. As of 1996, the U.S. Patent Office issued approximately forty new patents for mousetraps each year and rejected an unknown number of others.

The original 1903 mousetrap today accounts for more than half the mousetraps sold in the United States. But the mousetrap has evolved, and you can buy a mousetrap with a better latching device or a trap that is baited with an odorant instead of the traditional piece of cheese or blob of peanut butter. Behe is correct in his assertion that the spring-loaded mousetrap was intelligently designed. But it was not designed in one fell swoop, and preexisting parts were not simply added to it. To the contrary, the mousetrap evolved as one tinkerer after another borrowed and modified parts from earlier designs. In the same way as evolution designed the bacterial flagellum!

Behe argues that the bacterial flagellum consists of three well-matched parts: a rotor, a bearing, and a propeller. Take away any one, he contends, and the flagellum fails to operate. Thus, the flagellum is irreducibly complex and hence designed. It is, however, not fair to consider the instant removal of one whole part. The flagellum evolved gradually, and the constituents coevolved with it. If we wanted to reverse the process, we would have to make a series of gradual changes, not a single giant change. In reality, evolution cannot be reversed, but we can perform a thought experiment—just imagine recording a movie of the flagellum as it evolves, and playing the movie backward to watch it unevolve. The flagellum in the movie will not suddenly lose components one by one. Behe is not playing fair by demanding a massive change when a series of gradual changes is called for.

Behe's favorite analogy of irreducible complexity is the mousetrap. Take away any one part, says Behe, and the mousetrap fails to function. The analogy, however, is flawed for at least two reasons. First, mousetraps do not reproduce.

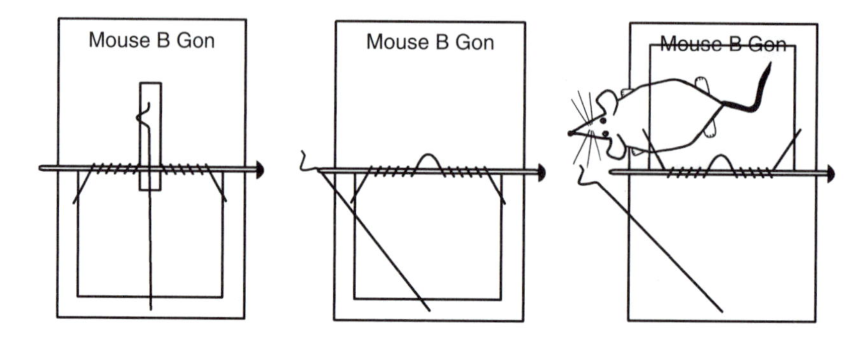


5. Mousetrap. The left figure shows a mousetrap with all its parts: latch, spring, trigger, hammer, and pin. In the central figure, the latch has been removed, but the trigger is held in place by the end of the pin. The right figure shows that the mousetrap still works, though it is not as effective, because it will trap only mice that approach from one side. Mouse courtesy of Shaunie and Ted Smathers.

They propagate by blueprints, not by genetics. Blueprints are drawings that provide exact specifications for a third party to follow in construction, and all mousetraps of a given generation are nominally identical. Any changes from one generation to the next are apt to be substantial.

The genetic code, however, is more like a recipe than a blueprint. Genes tell a leopard to have spots, not where each spot is located, just as yeast tells a cake to have bubbles, not where each bubble is located. Mice, unlike mousetraps, propagate by recipe, not by blueprint. Each mouse in a given generation is different from each other mouse. Changes from one generation to the next are apt to be small, but favorable changes may be subject to natural selection and their owners preferred over other mice. Organisms are not at all like mousetraps.

The second flaw in Behe's analogy is that the mousetrap is not irreducibly complex. We can remove several parts one by one, adjusting them as we do so, and still have a functioning mousetrap. Figure 5 shows a mousetrap whose latch has been removed. The metal bar that holds the hammer is now fixed in place by the central pin that holds the spring. The reduced mousetrap is not as good as the original, because it will not trap a mouse that takes the bait from the left side, but it still works. Other parts can be removed as well, but they have to be bent to make the trap work. (John H. McDonald of the University of Delaware has shown how to remove parts one by one and still make a working mousetrap. He has succeeded in reducing the mousetrap to two parts, but could not actually build his one-part design.)

Now imagine reversing the process and adding back the latch. Suppose that the latch had had some other function, and we borrowed and modified it to improve on the latchless mousetrap. Voilà! Evolution in action.

Behe presumes that the critical parts can have no other function and cannot change as they evolve. In fact, there are many kinds of flagellum, and not

IRREDUCIBLE COMPLEXITY DEMYSTIFIED

Hermann J. Muller, an American geneticist, won the Nobel Prize for his 1926 discovery that X-rays can cause mutations. In 1918, Muller discovered what Michael Behe later called irreducible complexity. But Muller recognized that such complexity was a commonplace result of evolution: "thus a complicated machine was gradually built up whose effective working was dependent upon the interlocking action of very numerous different elementary parts or factors, and many of *the characters and factors which, when new, were originally merely an asset finally became necessary* because other necessary characters and factors had subsequently become changed so as to be dependent on the former" [italics in original].

Behe thinks incorrectly that evolution adds parts, one by one, as if organelles were made out of whole cloth. Muller recognized that parts were gradually modified to perform new functions, until eventually one or more parts became crucial. In 1939, he coined the term "interlocking complexity," and thus anticipated Behe's "discovery" by nearly sixty years.

all are used for motion, so it is misleading to talk of "the" flagellum. Some flagella are used for parasitism, some for adhesion, and some for secretion. Similarly, some bacteria glide rather than swim. We do not know precisely how Behe's rotating flagellum evolved, but it is not unique and is related to flagella used for secretion. Very possibly a secretion system evolved into a system devoted to gliding motion on a film and ultimately into a rotating flagellum. We do not yet know the details, but we know enough about the relationships between different types of flagellum to devise plausible mechanisms whereby the flagellum used for swimming could have evolved from a less-complex flagellum used for other purposes. If half a flagellum is a secretion system, then half a flagellum is very useful indeed.

## THE EDGE OF EVOLUTION

In a later book, *The Edge of Evolution*, Behe shows clearly that he accepts common descent. He nevertheless constructs a Dembski-like probability argument to suggest that microevolution (here interpreted to mean changes within species) is possible but that evolution cannot develop entirely new forms or attributes. If that is so, then every innovation that ever developed, at least since the origin of multicellular organisms, must have been created by an intelligent designer.

Behe begins with the parasite *Plasmodium falciparum*, which causes malaria. In a comparatively short time, *Plasmodium* became resistant to chloroquine, the drug most commonly used to treat malaria. Chloroquine resistance requires two mutations, which Behe presumes have to be simultaneous. He notes that chloroquine resistance has appeared independently approximately ten times since the widespread use of chloroquine began. Making a rough estimate of the number of *Plasmodium* parasites that may have been exposed to chloroquine, Behe guesses that the probability of a single parasite's developing chloroquine resistance is approximately 1 in 100 billion billion ($10^{-20}$).

The biologist Nicholas Matzke of the University of California at Berkeley reviewed *The Edge of Evolution* in the journal *Trends in Ecology and Evolution*. Matzke argues that chloroquine resistance is more probable than Behe supposes and indeed that Behe's two mutations are not necessarily found in every resistant strain. To the contrary, Matzke thinks that "weak-but-selectable" resistance to chloroquine may be comparatively common. If that is so, then weakly resistant parasites may become prevalent and later gain in resistance when another mutation occurs. One mutation that gives partial resistance can prevail in a population, after which second and subsequent mutations grant additional resistance. There is thus no reason to believe that two mutations have to occur simultaneously in the same parasite.

Behe next presumes that any mutation or set of mutations less probable than chloroquine resistance is too improbable to occur by chance in multicellular organisms, because they simply do not have the large populations that *Plasmodium* exhibits. He further presumes, without supporting argument, that development of new forms would require a mutation whose probability is almost immeasurably small. That is, even though Behe accepts common descent and recognizes that successive body plans, for example, are built on preceding body plans, he thinks that a designer had to intervene to create teeth, hair, and mammary glands (not to mention, as Matzke points out, rattlesnake venom).

P. Z. Myers, of the University of Minnesota, on his well-known science blog *Pharyngula*, likens Behe's reasoning to that of someone who is studying poker and has discovered that the probability of drawing a royal flush is approximately 1 in 60,000. This person concludes, therefore, that a player can never win a game of poker, because that player will never draw a royal flush. The reasoning is faulty, obviously, because to win a round of poker, all you have to do is have a better hand than everyone else; sometimes a pair of threes will win. In the same way, major evolutionary changes do not necessarily require simultaneous mutations in the same organism. As we will see in chapter 10, evolution can be relatively gradual, and speciation need not take place in a single jump. Indeed, speciation is sometimes so gradual that the point at which one species is said to become another is entirely arbitrary.

### Tornado in a Junkyard

Dembski and Behe have, in essence, revived an argument called the *tornado in a junkyard*, which was promulgated by the astronomer Fred Hoyle. Hoyle argued that the development of a cell by chance was as likely as for a tornado to sweep through a junkyard and assemble a Boeing 747. Hoyle was right. But no competent evolutionary biologist thinks that a cell was assembled purely by chance. Neither, in a sense, was the 747: Airplanes began with attempts to emulate bird flight, progressed through the Wright brothers' plane at Kitty Hawk, through the *Spirit of St. Louis* (the airplane that Charles Lindbergh flew solo across the Atlantic Ocean in 1927), the Douglas DC-3 (the twin-engined airplane that revolutionized air travel in the late 1930s), and so on, to the Boeing 747. In the same way, biological organisms evolved by a series of gradual steps from the precursor to the first cell, through primitive single-celled organisms, to multicellular organisms. Complex biological systems such as cells did not develop suddenly by chance, and naïve probability arguments such as Hoyle's are wholly irrelevant.

The way in which a complex biological system takes the next step in its evolution depends on its history. Consider the following not-so-random sequence: FREDHOYLE. Suppose that we drew nine letters out of a hat filled with letters. The probability that the first letter was an *F*, for example, is independent of the value of any of the other letters. The probability of getting *F* in the first position is 1 in 26, presuming that the hat contains equal numbers of each letter in the alphabet. If the hat contains a very large number of letters, the probability of getting *R* in the second position is likewise 1 in 26; and so on. Since there are nine letters in the sequence FREDHOYLE, the probability of drawing that sequence FREDHOYLE in one try is approximately 1 in $5 \times 10^{12}$. Thus, it is extremely improbable, though not impossible, that we will draw the sequence FREDHOYLE in one try.

Suppose, though, that we draw the letters sequentially. If the first letter that we draw is *F*, then that value is fixed thereafter, and only eight letters remain to be drawn. The probability that the remaining letters will be RED HOYLE is increased by a factor of 26, to 1 in $1.5 \times 10^{11}$. If the next letter we draw is *R*, then the probability that the next seven letters will be EDHOYLE is increased by another factor of 26, to 1 in $8 \times 10^{9}$. We may continue in this way until we have drawn the first eight letters: FREDHOYL. The probability that we draw the last E and therefore the sequence FREDHOYLE is now a mere 1 in 26.

Let each letter represent a stage in the evolution of an organism. To incorporate selection into our example, we draw a lot of letters and throw out those we do not like. Possibly we employ many people to draw letters out of many hats. The probability of eventually drawing a letter we like goes way up. Once a sequence has evolved through the first eight stages, the probability that it will evolve to the ninth stage, represented by the letter *E*, is now substantial. Thus, drawing the sequence FREDHOYLE (or some other meaningful expression) is not at all improbable once many of the letters have been drawn. In the same way, development of a Boeing 747 or evolution of a multicellular organism is not so improbable, once we consider the previous construction of the Douglas DC-3 or the evolution of a single cell. The tornado in a junkyard is not a meaningful model of evolution.

### Information

A gene (genes are discussed in more detail in chapter 12) may be visualized as a sequence of only four letters, A, C, G, and T. Let us compare a gene to a sequence of letters that is wholly random. Any letter in that random sequence may appear in any location with equal probability. Since none of the letters is in any way constrained, the number of possible sequences of those letters is very large, and we say that the random sequence displays high *uncertainty*. (Uncertainty is closely related to thermodynamic entropy and is sometimes called *entropy*. In communication theory, it is sometimes called information, but here we define information as a reduction of uncertainty.) Without going into detail, we use uncertainty as a measure of the randomness of a sequence; high uncertainty implies a highly random sequence; low uncertainty implies a specific sequence. We will illustrate this principle using our FREDHOYLE example.

A gene is a sequence that contains what physicist Mark Perakh calls a *meaningful message* (in an apparent attempt to clarify Dembski's obfuscatory terms, specified complexity and complex specified information). FREDHOYLE is an example of a meaningful message. Think of it as a gene that codes for the person Fred Hoyle. If we change or delete some of the letters, we are apt to destroy the meaning or, indeed, change the meaning. REDHOYLE and EDHOYLE, for example, are genes that probably code for different people from Fred Hoyle. FREDHQYLE probably does not code for anyone. If we want our message to refer specifically to Fred Hoyle, then it must read FREDHOYLE and nothing else. We thus say that the sequence FREDHOYLE displays low uncertainty. Following the biologist Tom Schneider of the National Cancer Institute, we define the *information* in the sequence as the difference between the (high) uncertainty of a random sequence of nine letters and the (lower) uncertainty of the meaningful message FREDHOYLE. Copies of the gene FREDHOYLE are not considered to increase information, because they convey the same message.

Dembski argues that comparatively short genetic messages such as FREDHOYLE could easily have evolved during the time available since the origin of life, but that longer messages, messages that contain more than 500 *bits* of information (roughly the information on a single typed page), cannot have evolved within the available time. Consider, however, a long genetic message, analogous to FREDHOYLE but containing perhaps 300 or 400 bits of information. Suppose that the gene containing the message duplicates itself, so there are two FREDHOYLE messages. There has been no net increase of information, since both genes code for Fred Hoyle. But now one of the genes mutates and loses its first letter to become REDHOYLE. REDHOYLE no longer codes for Fred Hoyle, but rather codes for another person, Red Hoyle. If both genes contain 300 or 400 bits, then the new genome contains well over Dembski's limit of 500 bits.

There are many ways to exceed Dembski's 500-bit limit. A single-celled organism can acquire information by subsuming another organism and incorporating the genome of the second organism into itself; biologists think that mitochondria and chloroplasts originated as independent organisms that were absorbed by other single-celled organisms (see chapter 13). We jokingly call complexity attained in this way *agglomerated complexity*, by analogy with Dembski's term specified complexity.

## Conclusion

Dembski and Behe apply oversimplified or inappropriate analogies to the evolution of complex biological entities. Dembski thinks that evolution has a specific target, whereas, as we will see, it merely bumbles along looking for the best available solution to a specific problem. Evolution is blind and never finds the overall best solution; it is not goal-oriented. Behe thinks that a complex organelle such as a flagellum has to be assembled out of whole cloth from preexisting parts, whereas it is in reality assembled from a number of parts that evolve together.

Dembski and Behe are searching for gaps in our knowledge of evolution. They point to an organ or structure that is irreducibly complex and observe that we do not know how that organ has evolved. They segue seamlessly to the proclamation that we will never know, because God did it. Such reasoning is often called a *God-of-the-gaps* argument, because God is found in, or, inserted into, the gaps of our understanding. As gaps are filled by the advancement of science, however, such reasoning becomes more and more transparently flawed. We have a pretty good understanding of the evolution of the eye, and we are beginning to understand the flagellum. What gap will creationists exploit when the flagellum is fully understood?

In the next unit, "The Science of Evolution," we will look more closely at how evolution really happens. But first, we will take a quick romp through Charles Darwin's adult life and what led to his finally publishing his ideas on descent with modification and evolution by natural selection.

THE INVISIBLE HAND

Here (figure 6) is the solution to the problem posed in connection with figure 3: Why are the cannonballs not distributed uniformly about the target? Because, unknown to us, they were fired into a gully, and some of the cannonballs rolled toward the bottom of the gully. The gully represents side information, which is often unknown to us, even though it may be crucial to our understanding of a problem. The possibility of unknown side information completely nullifies the explanatory filter as a useful tool.

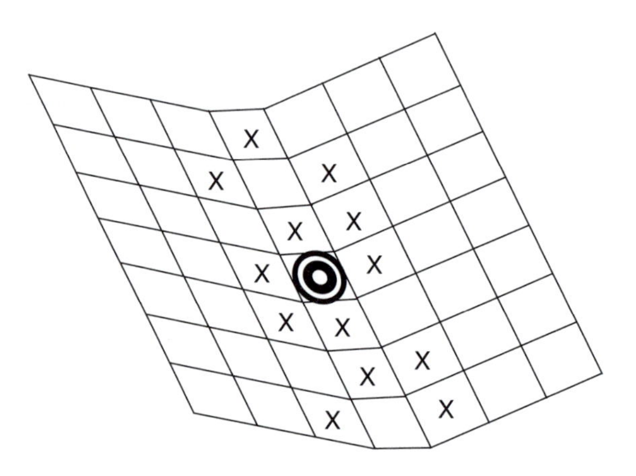


6. Solution to the problem posed in connection with figure 3. The cannonballs are not distributed uniformly around the target because they were fired into a gully, and some of the cannonballs rolled toward the bottom of the gully. The existence of the gully represents crucial side information that was unknown to us and invalidates the explanatory filter.

THOUGHT QUESTIONS

1. Should creationism (including intelligent-design creationism) be taught alongside evolution in public-school biology classes? Is doing so the same as teaching astrology alongside astronomy? If not, why not? Is it fair to ask high-school students to decide what is good science and what is not?
2. You are a science teacher. Your principal comes to you and tells you that you must now give equal time to teaching ideas you know to be untrue. How do you think you would react? Find out if your teachers or professors have ever been put in that position and, if so, what they did.

Part Three

# *The Science of Evolution*

Chapter 9

# *The Father of Evolution*

The fact of evolution is the backbone of biology.

—Charles Robert Darwin

When Charles Darwin was sixteen years old, his father decided to send him to school to study medicine, just as Darwin's father and grandfather had. Darwin lasted two years before his father pulled his easily distracted son out of medical school and enrolled him in the University of Cambridge to study for the clergy. After three years at Christ's College, Cambridge, Darwin graduated tenth out of 178 students with the intention of settling down somewhere as a country clergyman. While in school, however, Darwin had spent most of his free time taking courses on natural history from a Cambridge botany professor, John Henslow, and wandering the countryside collecting beetles.

Impressed with Darwin's innate talent as a naturalist, Henslow recommended him for a position as a naturalist on a mapping voyage of the southern hemisphere. Henslow convinced Darwin's father not only to grant permission but also to fund Darwin's portion of the trip on the vessel, the H.M.S. *Beagle*, which carried the surveyors and supplies. The nearly five-year journey was a life-changing event for Darwin. Three months after his return to England, Darwin was already beginning to formalize his ideas about biodiversity and geology. His first scientific presentation was to the Royal Geological Society in London, where he argued that South America, especially its western coast, has risen slowly over eons and that the local plant and animal populations have evolved adaptations as they found themselves in new and different environments.

It took Darwin only sixteen months to publish the first volume of *Zoology*, his descriptions of the animals he collected on his voyage, but Darwin waited almost twenty-two years after the *Beagle*'s return to England to make public his ideas on the evolution of species. The catalyst to finally publish his ideas was a paper sent to him by another naturalist, Alfred Russel Wallace, who was studying and collecting species in the Malay Archipelago. Wallace hoped that Darwin could comment on his ideas and help him decide whether they were worthy of publication. Wallace's paper was titled "On the Tendency of Varieties to Depart Indefinitely from the Original Type." Darwin was shocked to find that Wallace had arrived at some of the same conclusions about evolution by natural selection that Darwin himself had developed during and since his trip on the *Beagle*. While some of Wallace's ideas did not concur with his own, Darwin realized that he was now in direct competition with another scientist.

Darwin decided to present his and Wallace's ideas together in London at the July 1, 1858, meeting of the Linnean Society. Along with Wallace's current paper, Darwin presented two sections of his own nineteen-year-old manuscript to establish that, whereas both naturalists had arrived at similar conclusions, Darwin had begun formulating the ideas much earlier. A year and a half later, *On the Origin of Species by Means of Natural Selection* was finally published on November 22, 1859.

### IF DARWIN HAD A TIME MACHINE

Darwin and Wallace initially had no idea how descent with modification and evolution by natural selection really work. More specifically—and importantly—they had no knowledge of the raw material upon which the fitness of an individual is based: chromosomes, genes, and the molecules upon which genes are based. In fact, Darwin believed to his death that when two very different-looking parents produced offspring, traits were blended—as if the traits behaved like fluids. Darwin worried that, as a result of blending, any adaptations that allowed some individuals to produce more offspring than others could be swamped by chance mating with other individuals lacking those adaptive traits, thereby reducing variation in a population. Darwin also worried about novel structures, like the vertebrate eye, that seemed to be so improbable that their evolution through gradual selective processes seemed to him almost impossible (see chapter 8).

In 1866, Gregor Mendel published his paper describing experiments with pea plants and set forth his theory of *particulate inheritance*. Darwin may have been able to answer some of his questions about the mechanisms of inheritance had he read Mendel's paper, which was published in an obscure Czech journal. Historians have found no evidence (and many have tried in vain) that Darwin had any knowledge of Mendel's work in general and his results in particular. Yet the rediscovery of Mendel's research and conclusions at the turn of the twentieth century provided support for one of Darwin's arguments within his theory of evolution, that of common descent. Darwin's uncertainty about novel structures could have been eased had he known about evolution at the molecular level. Evolution is constantly remodeling the form and function of individuals within populations. This remodeling is possible because of the nature of genes and how and when they are expressed; this knowledge has become available only in recent years.

Today, as we pass the 150th anniversary of the publication of *On the Origin of Species*, we have compiled a massive, continuously and rapidly growing, and empirically based understanding of evolution. In fact, not a single scientific discovery in phylogenetics, paleontology, biochemistry, embryology, and development, to name a few, refutes Darwin's main hypotheses, namely descent with modification, which he put forth in 1859. To the contrary, research in these fields and in evolutionary biology itself has triumphantly confirmed Darwin's original theory of evolution by natural selection, as well as elucidated other mechanisms of evolution that Darwin could not have foreseen, such as *genetic drift* (see chapter 10), which is random where natural selection is deterministic. This mountain of support for evolution, in the form of tens of thousands of studies to test and/or refute it, and the absence of peer-reviewed scientific studies that falsify it have vaulted the theory of evolution to the most important concept in biology.

Furthermore, a number of opposing ideas that have been raised over the last 200 years, such as Lamarck's inheritance of acquired characteristics (see chapter 3), have been subjected to the same rigorous examination as Darwinian evolutionary theory and have been discarded because they were not consistent with actual observations. Hence, modern evolutionary theory is almost universally accepted, but not because of a dogmatic support since the time of Darwin. To the contrary, it has emerged from among a number of ideas as the only one that is supported by the facts. At times in the past, groups of scientists have only grudgingly accepted the ideas that make up evolutionary theory, but as the evidence grew, the modern theory of evolution triumphed.

The theory of evolution itself has evolved since Darwin first put forth his theory in 1859. Evolution as we understand it today includes at least sixteen fundamental principles, all of which are outlined in the box "The Fundamentals of Evolution."

## Conclusion

While Charles Darwin was not the first naturalist to propose the evolutionary principle that species change through time, he was the first to (1) propose a theory of evolution (descent with modification) based on observable and testable evidence, (2) provide a mechanism to explain descent with modification, and (3) synthesize his theory in a book available to the general public. Darwin collected some of his first evidence while traveling the southern hemisphere on the H.M.S. *Beagle*, but even after this trip, he continued to study evidence for evolution in species ranging from barnacles to pigeons. By publishing *On the Origin of Species*, Darwin focused the attention of the world's naturalists on the incredible diversity found among species, both past and

### The Fundamentals of Evolution

Douglas Futuyma of Stony Brook University is the author of dozens of articles and several books and textbooks on evolution; he has distilled the fundamentals of evolution into sixteen categories. We summarize them below:

1. The phenotype (observed characteristic) is different from the genotype (an individual's set of genes).
2. Acquired characteristics are not inherited.
3. Genes do not blend with other genes; they retain their identity as they are passed from one generation to the next.
4. Genes mutate.
5. Populations, not individuals, evolve as one genotype replaces other genotypes over generations.
6. The fraction of the genotype in the population changes either randomly (genetic drift) and/or by natural selection.
7. Even slight environmental change can promote significant evolutionary change.
8. Genetic recombination through sexual reproduction increases the variation among individuals in a population.
9. There is genetic variation in natural populations.
10. Geographically separated populations of a single species are genetically distinct.
11. Slight phenotypic differences between species or between populations of the same species are often a result of differences between several genes.
12. These phenotypic differences arise by natural selection and are often adaptive.
13. Biological species are defined as groups of genetically and phenotypically distinct individuals that interbreed with each other and do not exchange genes with other groups.
14. Speciation occurs when two populations become reproductively isolated from each other (usually by way of geographic separation) and those populations do not share genetic changes.
15. Small differences between two diverging populations can accumulate over long times until the two groups eventually are classified into different families, classes, phyla, or even kingdoms.
16. While the fossil record is still incomplete and thus contains gaps, it also contains evidence of gradual change from ancestral forms to descendants with considerable phenotypic differences.

present, and (jointly with Wallace) proposed a mechanism to explain that diversity. Darwin's theory has held up over and over again as scientists have tested it using everything from the fossil record to molecular diversity. All things considered, we therefore honor Charles Darwin with the title, the father of evolution.

In the chapters that follow, we will explain how evolution works and how we know it works, focusing on some, but not all, fields of biology that provide us with evidence in favor of descent with modification, including embryological development, genetics, the classification science of phylogenetics, and paleontology.

### Thought Questions

1. Why did Darwin wait twenty-two years to finally publish *On the Origin of Species*? Are things different now? What would you have done in his place?
2. Of the sixteen fundamentals of evolution put forth in the box, which do you think Darwin would have difficulty understanding and why?

CHAPTER 10

# *How Evolution Works*

Nothing in biology makes sense except in the light of evolution.

—Theodosius Dobzhansky, pioneering geneticist and evolutionary biologist

CREATIONISTS LIKE TO stress chance, as if chance automatically ruled out evolutionary change. Chance is indeed a factor in evolution, but it is not the whole story. Indeed, chance can often lead to very predictable behavior. Thus, before we go on to discuss evolution, let us examine a physical phenomenon that, like evolution, is governed by chance but leads to very lawlike behavior: diffusion.

Suppose that we prepare a shallow dish and fill it with water. In the dead center of the dish we inject a small droplet of dye (figure 7, left). With time, the droplet will spread out, and the color will become less intense (figure 7, center, right). Eventually, the dye will uniformly fill the entire dish. From that time on, the appearance of the solution will not change, assuming that we do not let the water evaporate.

How does the dye come to fill the dish? By a process known as *diffusion*. Diffusion is the result of random molecular motions—chance. Each molecule of the dye is constantly buffeted from all directions by the water molecules

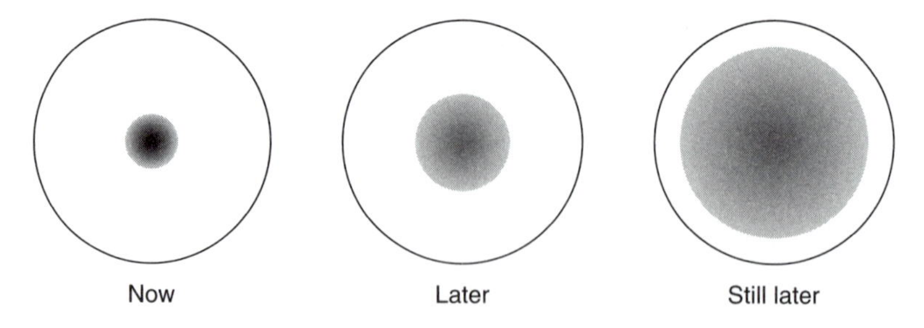


7. A droplet of dye injected into the center of the dish gradually diffuses outward until the dye fills the whole dish uniformly.

(and other dye molecules, but there are many fewer of them). Its path is therefore completely random—determined by chance.

Figure 8 shows the paths of three hypothetical dye molecules. Each follows a random path, but after a great many steps, the odds that any one will remain near the center of the original droplet are very small. Suppose that we repeat the experiment with the same three molecules (a physical impossibility, but this is a thought experiment). On the second, third, fourth, fifth, . . . experiments, they will follow very different paths from those shown in figure 8. It is impossible to predict where they will end up after a few minutes, but few will remain stationary, and all will be found a measurable distance from the center of the dish.

There are not three but billions of dye molecules. Instead of repeating the experiment with three molecules over and over, let us do it once with three *billion* molecules. Each follows a random path, but, on average, each drifts away from the center of the dish just as did the three molecules in figure 8. The result is what we have already seen in figure 7 and is highly repeatable. That is, if we know the temperature, the composition of the dye, and the liquid in which it is dissolved, we can measure the rate at which the droplet spreads out, or *diffuses*. If we repeat the experiment another time at the same temperature, we will find the same rate with very good accuracy. Thus, the rate of diffusion is highly predictable, even though it is the result of myriad random collisions.

Diffusion is a well-known and indeed important physical process that is a result of chance—random collisions—but results in very lawlike behavior. So is evolution. For example, in a small breeding population a rare, yet important *allele* (form of a gene) may disappear from the population if the individual or

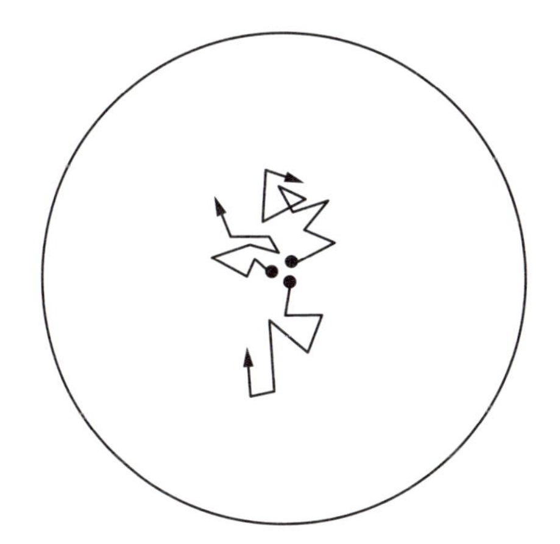

8. The paths of three hypothetical dye molecules. The chance that any one will remain near the center of the original droplet is small.

individuals carrying it fail to breed. This failure could be the result of a chance accident or disease. While this instance of genetic drift is random and due to chance, the probability of its happening in this small population is highly predictable.

### Diversity

When a teacher gazes out over her classroom of fifteen-year-old students, she is aware of the diversity in that small population. She first notices the obvious diversities, like sex and ethnic diversity. Upon further analysis, she begins to notice diversity in hair color and degree of curliness, diversity in height and weight, and diversity in face shape and facial features. Especially because her students are teenagers, she may even begin to notice diversity in the timing of physical maturation. We can easily notice diversity among humans because we are so familiar with (and likely focused on) subtle differences among individuals.

What if we replaced this teacher's students with a natural population of red-winged blackbirds? Could she detect diversity among individuals beyond the obvious gender difference between the brown-striped females and the black males with red patches on their shoulders? Would she notice differences within sexes, such as the stripe pattern and beak depth of the females, or differences in song or the size and brightness of the shoulder patches on the males? How about wing length and symmetry, aggressiveness, age, sperm count, or skills at raising babies? Experiments in mate selection by females show that blackbirds can detect subtle differences among individuals in their populations, even though we cannot distinguish them. Also invisible to us, and probably to the birds as well, is genetic diversity, such as a recessive gene that may produce infertile offspring.

Diversity in natural populations is a phenomenon of which most nonbiologists are unaware. Yet it is diversity at all levels—genetic diversity, diversity among individuals in a population, diversity among populations of a species—that allows a species to carry on through time. For example, if long wings give birds a slight survival advantage over short wings in a given year, then the long-wing trait will increase in frequency in the next generation of offspring. If the next year's conditions favor short wings, however, then that trait will increase in frequency while the long-wing trait decreases. Maintaining wing diversity in the population gives the species as a whole a survival advantage over long times.

The combination of all of the traits that vary among individuals in a population may allow some individuals to be more successful at surviving and producing offspring than other individuals. This survival and reproductive advantage is a product of evolution by natural selection and descent with modification.

Fitness

The smallest shrews are called least shrews, and are among the smallest mammals. They are about 7 centimeters long and weigh 5 or 6 grams. They have small, spindly legs. The elephant, by contrast, has strong, stout legs, which are necessary to support its weight. If we scaled a shrew to the size of an elephant, its legs would collapse unless we also altered its body plan substantially, in particular, making its legs stronger. This is so because the weight of the animal would increase more than the cross-sectional area of its legs. Similarly, if we scaled an elephant down to the size of a shrew, it would be virtually immobile because it would not be strong enough to move its (relatively) heavy legs. Each species has an optimal weight, given its environment, its means of gaining sustenance, and its physical characteristics.

*Fitness* means the ability to survive and reproduce; it is measured by the number of surviving offspring that an organism can produce. A shrew that is slightly larger and therefore heavier than the average might be somewhat less fit because it cannot avoid predators as well as its smaller, nimbler relatives. Similarly, a shrew that is slightly smaller than the average might be less fit because it is not strong enough to capture certain kinds of prey. Both shrews are fit, but less fit than they could be. On average, such shrews will produce fewer offspring or less-fit offspring than their fitter cousins.

We may describe this situation by drawing a series of graphs such as that shown in figure 9. The figure plots fitness as a function of weight. The fittest shrew has the optimum weight; any other shrews are at least slightly less fit.

Suppose that the average parents *P* of one population of shrews are somewhat lighter than the optimum weight. The fitness of *P* is represented by the letter on the left edge of the curve in the first panel. If weight is a heritable

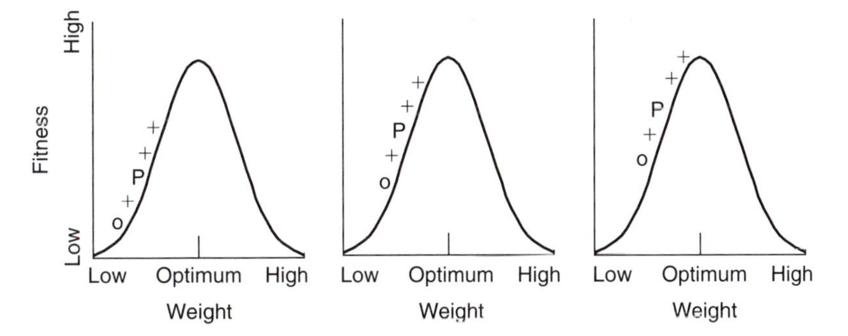


9. Successive generations of shrews climbing a fitness function. In each generation, *P* is the average parent, the + signs represent offspring, and the *O* represents an offspring that does not survive to reproduce. For convenience, we show four offspring, the least fit of which always fails to reproduce. With each generation, the average parent shows increasing fitness.

characteristic, then a light shrew will in general have lighter-than-average offspring. Some of these will be too light and will not survive to reproduce. Some, however, will be heavier than the parents and therefore on average more fit; they will produce fitter offspring, and more of them, than either their parents or their less-fit siblings.

The offspring of the parents *P* are shown as + and O: The offspring labeled + survive and reproduce, whereas the offspring labeled O do not survive. For convenience, we show four offspring, the least fit of which does not survive. The average fitness of the three survivors indicated by the + signs is now higher than that of the parents *P*. That is, the next generation of offspring in the population will, on average, be closer to the peak of the fitness graph than were their parents, as shown by the location of the *P* in the second panel. The process repeats, as in the third panel, and gradually the fitness of the average shrew increases until the average parent's fitness equals the optimum fitness; that is, *P* moves to the top of the curve. We could have made the same argument had the original parents *P* been heavier than the optimum weight.

The average parents move up the curve in a way that is similar to diffusion. Imagine that we are on the side of a hill, about where the first *P* is seen in figure 9. We take a handful of stones and toss them straight up. Some roll down the hill and are lost to us; these are represented by the *O* in figure 9. Others lodge in crevices somewhat uphill from where we are now; these are represented by the + signs in figure 9. We walk up the hill and gather these stones, as well as some others, and again toss them straight up. We repeat the process until we have arrived, by diffusion, at the top of the hill. We can go no higher.

If the environment does not change, on average, the graph will be about the same from generation to generation, because very heavy and very light shrews either do not live to reproduce or produce at least some offspring that are fitter than they.

Now, imagine that something in the environment of the shrew population changes abruptly and makes heavier individuals more fit. Shrews are insectivores, or insect eaters, and perhaps a drought has decimated some of their insect prey, leaving only larger prey for consumption. Smaller shrews are now at a disadvantage. We may describe the change by drawing another fitness function, which has a peak at a higher weight (to the right in the graphs). If the environment changes too much, then the fitness function may shift so far to the right that none of the present generation of shrews is fit enough to thrive and reproduce. This population of shrews therefore goes extinct, a fate that has befallen an overwhelming majority (99 percent) of all species that have ever lived.

Let us now suppose that the environment changes appreciably, but this time the new fitness function overlaps the old fitness function. Such a change can happen, for example, if a river carves a boundary that separates one group of shrews from the main population. Let us say that heavier shrews in the new population are now more fit than lighter shrews. The average shrews in the new environment do not have the optimum weight for that environment. Their offspring, however, show a range of weights, and the heavier of those are more fit than the lighter. In consequence, subsequent generations of shrews become more and more fit, until some shrews attain fitness near the new peak, precisely as shown in figure 9.

It is hard to define a species, but one common definition is that two populations are different species if they do not normally interbreed or cannot interbreed; evolutionary biologists call this condition *reproductive isolation.* Eventually, the new, heavier population of shrews may not be able to interbreed with the population across the river, even if they are somehow reunited (think of a Great Dane and a Chihuahua; although these are generally considered breeds, not species, they could reasonably be considered different species because they cannot normally interbreed. Indeed, the Great Dane and the Chihuahua exemplify the difficulty of precisely defining a species).

Even if the shrews from opposite sides of the river can interbreed, the offspring may not be fit enough to reproduce in either environment. Individuals from the two populations also may have become so genetically different due to the buildup of random mutations that their offspring, if any, may be sterile or too frail to produce their own subsequent offspring. This condition is called *outbreeding depression*, in which the offspring are essentially hybrids of the now very different parents. When the new population cannot interbreed with the original, we say that a new species has developed.

Creationists like to say that *microevolution* is possible, but that *macroevolution* is not. That is, they admit to minor genetic changes, or microevolution, within a species or, in their terms, a kind; but they deny that *speciation*, or macroevolution, is possible. The distinction between microevolution and macroevolution is to a large extent a false dichotomy; sometimes two species are so similar that only an expert can distinguish them; in that case, speciation *is* microevolution. At other times, however, there are substantial differences between closely related species.

Philosophers like to talk about the *heap problem*. One grain of wheat is not a heap of wheat; two grains of wheat are not a heap; but a million grains of wheat are indeed a heap. At what point did the number of grains become sufficient to be considered a heap? Ten? One hundred? One thousand? Obviously, at no specific point, yet somewhere the heap became a heap.

As we add grains of wheat, then, the collection of grains evolves into a heap. Precisely when it becomes a heap is both undefined and, in practice, undefinable. Yet, when it contains enough grains, the heap is unambiguously a heap. In the same way, speciation is no more than small genetic changes—microevolution—accumulating over time until we may consider that a new species has evolved. Precisely when a new species evolves is indefinable, but that is a trivial fact and not an argument against speciation, as we can see by considering the heap problem.

We have drawn the fitness function as a two-dimensional graph. In reality, weight is not the only factor that determines fitness, and we would have to draw myriad such graphs to represent all of the possible factors. In that case, we say that the fitness function is *multidimensional*. Nevertheless, the principle we described above still holds, and we find shrews with a range of characteristics, from weight to coloring to speed to sense of smell to intelligence. In the multidimensional case, the fittest shrew is the shrew that is fittest in a variety of attributes. Other shrews display a range of fitness in all these attributes. Because the fitness function is multidimensional, incidentally, a population need not be trapped forever at the peak of the fitness function; if a shrew cannot evolve better eyesight, then perhaps it can evolve better camouflage or stronger legs.

Importantly, the peak of the fitness function does not represent the overall fittest individual. Possibly the fitness function includes one or more peaks that have higher fitness but are not accessible to the present population of shrews. For example, a shrew might be fitter if it had eyes in the back of its head, but there is no known developmental process that is likely to outfit a shrew with four eyes evenly spaced around the head. As we will see, evolution works with the raw materials available to it and often finds solutions that are not optimal. Such suboptimal solutions may be considered evidence in favor of evolution and against an intelligent designer.

Finally, we want to stress that evolution has no predetermined direction. The characteristics in a population today are a product of selection pressures of the past. Populations constantly play a game of catch-up with changes in the environment. We often speak of an *evolutionary arms race* between predator and prey or parasite and host; for example, as the prey evolves better camouflage, the predator evolves better vision. It is not possible to predict over the long term where the arms race will lead.

### Dembski's Lunch

We drew the fitness function in the previous section as a smooth curve with only one major peak. The fitness function for a population of shrews might consist of two or more widely separated and nonoverlapping peaks, but then we would say that the shrews were two or more distinct species, each of which was described by only one of the peaks. The fitness function for a given species has a single peak and is fairly smooth.

Mathematically, however, we could imagine an infinite number of fitness functions. Some would be realistic, like figure 9, whereas others would be

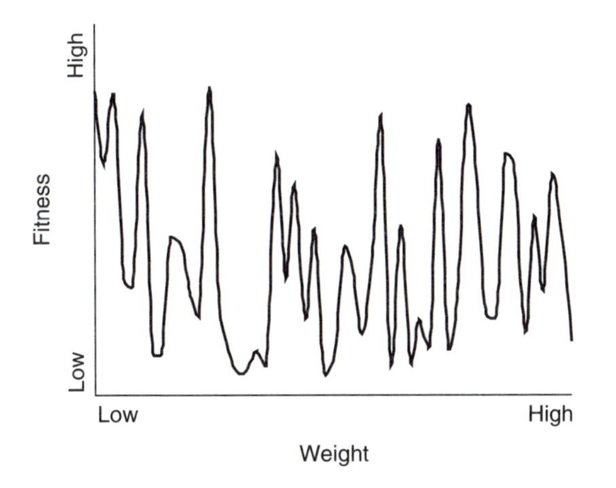


10. Random fitness function. A majority of mathematical fitness functions is rough, whereas real fitness functions are smooth. Note the regions of low fitness that the species cannot cross.

simply random jumbles of peaks, as in figure 10, which was drawn using a random-number generator. The random jumbles of peaks have mathematical interest, but they are not useful to biologists studying evolution.

Mathematicians have developed computer programs that model evolution as we have described it. These programs are called *evolutionary algorithms* (an *algorithm* is a step-by-step procedure for solving a mathematical problem, but the word is most often applied to a computational procedure carried out on a computer). Just as our population of heavy shrews climbed a hill generation by generation to attain maximum fitness, evolutionary algorithms may introduce random variations into a design, select the best designs of the next generation, and also climb a hill in a search for an optimum design. Evolutionary algorithms that work in this way are called *hill-climbing algorithms*. Such algorithms have been used for designing structures, such as camera lenses, that are too complex to design any other way.

A hill-climbing algorithm is a good heuristic device (a *heuristic* is a formulation that is not necessarily rigorous but provides a guide to solving a problem). Evolution, however, does not work exactly like a hill-climbing algorithm, for at least two reasons. First, even if every trait has its own fitness function, some traits are linked, either genetically or developmentally, or by the environment. It would therefore be very difficult, if not impossible, to define an average fitness function. Even if we could define such an average, there would be no reason to believe that natural selection operates equally on every trait that composes the average fitness function. Natural selection thus does not maximize the average fitness function. Rather, it operates on some traits under some conditions and on other traits under other conditions. Few if any traits are ever optimized; that is, few if any traits ever achieve the peak of their respective fitness functions.

Additionally, the fitness function is not fixed but responds as the organism improves its fitness. Thus, the lion gets faster as the gazelle gets faster, and the eagle gets sharper-eyed as the rabbit achieves better camouflage. The interaction between the organism and its environment is known as *coevolution*; coevolution cannot be described as a simple hill-climbing algorithm.

William Dembski, the inventor of the explanatory filter (see chapter 8), has nevertheless attacked the theory of evolution by applying some mathematical theorems that relate to evolutionary algorithms. The details are complicated, but the theorems prove that no algorithm is better than a random search *when averaged over all possible fitness functions*, including those that are not physically realizable and are therefore irrelevant to evolution. The theorems are called the *no-free-lunch* (NFL) theorems, and Dembski has accordingly called his book *No Free Lunch*.

In hindsight, the NFL theorems are unsurprising. Imagine that we wanted to find the highest peak in figure 10. How would we go about it? We would not use a hill-climbing algorithm, because there are no hills. We might try jumping from peak to peak, but how far would we jump each time? The distances between the peaks are random, so we have no way of knowing. We might as well throw darts randomly as wrack our brain trying to find a better way.

An infinite number of fitness functions is a lot, but that's how many mathematically possible fitness functions exist. Many are nice, smooth functions with a single peak, but most are not. Many are smooth functions with many peaks, and many are random jumbles (called *discontinuous functions*). When we average over all those functions, we find that every possible strategy for locating a peak is as good as any other.

But why do we want to average over all mathematically possible fitness functions? In biology, we don't. We want to know about biologically realizable fitness functions, and these are not random jumbles but nice, smooth functions. If we want to start at the bottom and find a peak, a hill-climbing algorithm will do very nicely—and substantially better than a random search, which will spend a lot of its time very far from its intended peak.

As the physicist Mark Perakh has pointed out in the book *Why Intelligent Design Fails*, Dembski has introduced random fitness functions and misused the NFL theorems to suggest that an evolutionary algorithm is no better than a random search *on a specific fitness function*. Evolution by a random search would indeed be impossible; a search has to be relatively gradual and continuous. Dembski uses his misinterpretation of the NFL theorems to "prove" that evolution is impossible because an evolutionary algorithm is no better than a random search. His faulty logic is as follows: Evolution can be described mathematically by an evolutionary algorithm; an evolutionary algorithm is no better than a random search; evolution cannot progress by a random search; therefore evolution cannot progress, period. The second step in this "logic" is incorrect because the theorems do not apply to a single, smooth fitness function but only to average overall fitness functions, nice as well as nasty. In biology, fortunately, we have mostly smooth fitness functions. Natural selection is not ruled out by the no-free-lunch theorems. Indeed, natural selection is so well established that, if the no-free-lunch theorems had cast doubt on it, we would have had to question their applicability to evolution, rather than reject evolution.

### As American as Apple Maggots

The apple is an import from Europe. The North American apple maggot is the larva of a certain kind of fly. It today feeds on apples, but it originally fed on the fruit of the hawthorn tree and adapted to the apple only after the importation of the apple from Europe. Entomologists asked whether the flies had speciated into hawthorn flies and apple flies. The flies are difficult to distinguish, so the scientists dyed them and watched them for hours, ultimately concluding that apple flies mated almost exclusively with other apple flies, whereas hawthorn flies mated with hawthorn flies. Only a small percentage of the flies cross-bred.

The two cohorts of fruit flies no longer interbreed because apples and hawthorns set fruit at different times, and the maggots hatch to coincide with the setting of the fruit. This form of reproductive isolation is called *temporal isolation* (isolation in time) because the reproductive schedules of the two fly populations do not overlap. Yet other flies are closely related to the apple flies and have the same geographical range but feed instead on blueberries. The hawthorn and blueberry fly species, which are now clearly distinct, may have diverged after adapting to specific plants. Thus, we are clearly seeing the divergence between apple and hawthorn flies and may have seen, in the blueberry flies, an actual example of speciation.

What we have just described is only one mechanism whereby evolution takes place. The change in the environment here was the introduction of the apple tree. Other changes can include the appearance or disappearance of a predator or a disease, a change of climate, or the appearance of a new river that cuts one population off from another. Traits that are not related to fitness may also evolve by accidental genetic changes, or *genetic drift*, or by *sexual selection*, as when females preferentially choose males with certain traits. For example, in less than four million years, crickets in the genus *Laupala* in the Hawaiian archipelago have evolved into thirty-eight distinct species. All thirty-eight species are ecologically indistinguishable (same habitats, same food source, and so on) except for the specific mating call of the males of each species, which

### Turnabout

Thrift is a virtue, right? Not necessarily. Not when we see it in a member of a minority group. Then it is not thrift but stinginess.

At least that's the way many prejudiced people see it. An attribute that they consider a virtue in themselves is miraculously transformed into an evil in people they dislike. Thus, "I spend my money wisely; Those People (whoever they may be) are stingy." "I have a large circle of friends, and we all belong to the same church; Those People are cliquish." "I like to relax and enjoy my coffee breaks; Those People are lazy."

According to William Dembski, Phillip Johnson, and others, supporters of evolution know full well that the theory is in tatters, but they support it because of their dogmatic adherence to naturalism. We argue, to the contrary, that Dembski and Johnson oppose evolutionary biology, in spite of the evidence and the scientific consensus in its favor, because of their own religious dogmatism. In effect, they see stubbornness and dogmatism as virtues in themselves but not in their opponents.

We would say, rather, that biologists are firm in their convictions, whereas creationists are being dogmatic, holding unreasonably to their preconceived notions. The theory of evolution is based on observation, theory, evidence, and inference. William Dembski's statement, "As Christians, we know naturalism is false," is not even an inference. It is a flat statement that not all Christians agree with and has not a shred of supporting evidence. It is dogma. And dogma has no place in science.

is preferred by the females of that species only. The long tails of peacocks and other birds are probably also the result of sexual selection.

## Conclusion

Evolution works in large part by exploiting variations among individuals in a population and changes in the environment. It is conceptually easy to understand: Individuals of each sex breed and produce offspring with a range of fitness. If the environment changes, then a population of highly fit organisms may suddenly, on average, become substantially less fit relative to the new environment. The most-fit individuals among these survive in greater proportion, and/or produce more offspring than the least fit (*differential reproductive success*). Thus, certain pressures in the environment have, by chance, selected these individuals and their unique characteristics. The average fitness of a population must trend toward optimum if the population is to continue to exist. As the characteristics of the population change over time in response to conditions and the population reaches a point at which it can no longer interbreed with the ancestral population, we conclude that a new species has evolved. Speciation, as we have described with the apple maggot fly, has been observed in the field, not just in the fossil record.

Evolution may be described graphically by a fitness function that looks, in one dimension, like a hill. Procedures for simulating evolution are sometimes called hill-climbing algorithms. Hill-climbing algorithms work because realistic fitness functions are smooth. Attempts to discredit the theory of evolution by appeal to probabilistic arguments or the no-free-lunch theorems are simplistic and flawed. There is no debate among scientists whether evolution (descent with modification) takes place, nor whether the theory of evolution accurately accounts for descent with modification: It does. The debates concern the precise mechanisms. For example, does speciation occur primarily in small, isolated populations, or is it a broader, general phenomenon? Though biologists have opinions, no one really knows. Similarly, does speciation occur relatively quickly, with long periods of intermittent stability (*punctuated equilibrium*), or is it a slow, gradual phenomenon? Or is it slow in some species and punctuated in others? Again, biologists have their preferences, but no one knows for sure.

Scientific debates or differences of opinion such as these have allowed creationists to pretend that evolution is a theory in crisis, propped up by dogmatic supporters who are virtually religious in their defense of their failed theory. As in all areas of science, these debates in fact help to solidify some aspect of evolution, while they allow us to discard old, falsified hypotheses of evolutionary processes and replace them with new explanations.

In the next chapter, we discuss how biologist Ernst Haeckel's overenthusiasm for Darwin's theory of evolution led to a century of controversy, misleading textbook illustrations, and a certain amount of misinformation. Specifically, we discuss Haeckel's misinterpretation of the very real similarities among vertebrate embryos, which led to decades of textbooks published with a misleading drawing. Finally, we discuss the concept of recapitulation and how it helps us understand why there are so many structural similarities among vertebrate embryos.

## Thought Questions

1. Creationists sometimes argue that their belief is as valid as a belief in evolution, that they simply have a different starting point from scientists. Creationists believe in the Bible (or other scripture) whereas scientists depend on observation and experimentation. Thus, say creationists, evolution is a religion. Is it?
2. California requires a course in biology for entrance into the state college and university system. Is that a reasonable requirement? Can it be

properly construed as religious discrimination if someone does not want to study biology for religious reasons?

3. Can you think of any traits in humans or other animals that may be attributed to sexual selection? Why do you think they are the result of sexual selection? How might they have been selected?
4. Why do you think that biological fitness functions have to be smooth? What might happen if they were random jumbles (discontinuous functions)?

CHAPTER 11

# *Recapitulation*

I myself am convinced that the theory of evolution, especially the extent to which it's been applied, will be one of the great jokes in the history books of the future. Posterity will marvel that so very flimsy and dubious an hypothesis could be accepted with the incredible credulity that it has.

—Malcolm Muggeridge, British journalist and author

IN 1874, A GERMAN comparative embryologist and Darwin enthusiast, Ernst Haeckel, began publishing his drawings of vertebrate embryos in various stages of development to explain common ancestry and to support Darwin's theory of evolution. Haeckel drew the first phylogenies and coined the term "tree of life" (see chapter 13). His skillful and detailed drawings clearly showed that the early-stage embryos of many species of vertebrates are nearly if not wholly identical. With his drawings and accompanying descriptions, Haeckel promoted the idea that "ontogeny recapitulates phylogeny," also called the *biogenic law*. The biogenic law states that the path of an organism during its embryological development (ontogeny) is a summary of its evolutionary history (phylogeny). For example, a human embryo has structures that resemble gill slits, just like a fish. The human embryo then loses these structures and grows a tail and four limbs, resembling a reptile. The tail disappears, and the embryo begins resembling a primate. Haeckel tirelessly promoted his biogenic law, and many biologists in the nineteenth and into the twentieth centuries embraced it as a simple explanation of evolution. How convenient that all we had to do was look at embryological development to understand an organism's evolutionary history!

While Haeckel had many supporters, he was not without his early critics. Almost immediately, many of Haeckel's fellow embryologists noticed that he had taken artistic liberties in his drawings to support his ideas, yet few rejected the biogenic law outright. One of the first written criticisms of the biogenic law appeared in an 1894 article by zoologist Adam Sedgwick in the *Quarterly Journal of Microscopic Science*. Sedgwick argued that the biogenic law conflicted with a principle known as *von Baer's law*, after Karl Ernst von Baer, one of the founders of embryology. Baer had noted, contrary to Haeckel, that the embryos of higher animal forms resembled the embryos, not the adults, of earlier forms. Thus, for example, a human embryo may pass through a stage in which it resembles a fish embryo, but not an adult fish. Sedgwick noted, "Embryos of different members of the same group are more alike than the adults, and the resemblances are greater the younger the embryos examined." He continued that when the actual embryos were examined, "a blind man could distinguish between them."

As the twentieth century unfolded and the fields of empirical embryology and genetics emerged, it became clear that in his drawings Haeckel had emphasized similarities between the embryos of various vertebrate classes and neglected the differences. It is not clear whether he purposely altered his drawings to better fit his ideas. In any case, the drawings fascinated lay people and scientists outside the fields of embryology and evolution. Biology textbook authors looking for detailed and authoritative images to illustrate their chapter pages happily included the drawings. Due to the cost-effective practice of recycling images and their accompanying explanations, as well as the paucity or even lack of evolutionary biologists on the editorial staffs of textbook companies, the drawings remained in many textbooks until the 1970s. During this time, many teachers who had little or no background in evolution taught their students that the drawings were evidence of evolution, and the students were encouraged to understand and memorize the catchy phrase, "ontogeny recapitulates phylogeny."

Finally, in 1977, in his book *Ontogeny and Phylogeny*, Stephen Jay Gould carefully dissected Haeckel's drawings and disproved the general ideas behind the biogenic law. Later, in one of his last essays in *Natural History* magazine (March 2000), Gould explained that while Ernst Haeckel was regarded among his contemporaries as a master naturalist, he often "took systematic license in 'improving' his specimens to make them more symmetrical or more beautiful." Most likely as a result of Gould's careful critique, Haeckel's drawings are no longer found in today's biology textbooks.

Haeckel was not wholly wrong. As we will see later, phylogeny is indeed to some extent recapitulated in embryological development. What Haeckel got wrong, however, was his belief in the recapitulation of the adult forms. Rather, embryological development of today's vertebrates summarizes the evolution of past embryos. Haeckel's conception that evolution is inherently progressive, however, is no longer accepted. Whether Haeckel deliberately fudged his drawings remains a matter of debate. Unfortunately, the creationist Jonathan Wells, in the book *Icons of Evolution* (see chapter 3), exaggerates the importance of Haeckel and the biogenic law on the theory of evolution, and purports to show that the entire theory of evolution is founded on a handful of errors such as the biogenic law. In reality, as we have seen, biologists disputed the biogenic law early on and eventually replaced it with the more-realistic von Baer's law. The eventual discovery of Haeckel's misstep is an excellent example of how science self-corrects. That such mistakes are uncovered is not a weakness of science, but a strength. Haeckel's drawings were mere bumps on the road to a comprehensive theory of evolution.

### Recapitulation

In its original form, the idea that "ontogeny recapitulates phylogeny" is not an accurate evolutionary statement. Recapitulation nevertheless provides helpful insight into evolutionary relationships and ancestry. Harvard University zoologist Ernst Mayr described recapitulation in embryological development as the appearance of an ancestral structure found in two different lineages, say, for example, the pharyngeal arches and pouches (see below) that are found in the embryos of both fish and mammals. The structure then disappears from the embryo (or provides the organizational foundation for future structures) in one lineage, mammals, but is maintained in the adult form of the other lineage, fish. Mayr argued that this structure provides evidence that these two lineages are, in fact, connected to a *common ancestor*, that is, a group of organisms, not a single individual, from which two lineages most likely descended.

But does not an organism waste energy on a structure that will disappear later during embryological development? To the contrary, these ephemeral ancestral features provide a framework upon which the successful development of future structures depends. In other words, these ancestral characteristics are now playing new roles in embryological development, like organizing tissues that will eventually become bones in the skeleton of an individual. For example, the *notochord* is a character that provides internal structural support and unites all members of the phylum Chordata (or chordates, animals with notochords and pharyngeal arches, among other characteristics). At some point in development, every species within Chordata, from sea squirts to sea lions, possesses a notochord. Most members of the chordate phylum, those in the subphylum Vertebrata, replace the notochord with vertebrae (the backbone) later in development to support the pelvic and pectoral girdles (to which the front and hind limbs are attached) and to protect the spinal cord. Yet in vertebrates, as the notochord is forming (during a developmental process called *gastrulation*), it first functions to establish the midline of the embryo. Once the notochord forms, its cells send molecular signals to neighboring cells, inducing those cells to begin developing divisions of the central nervous system and eventually the brain and spinal cord. Thus, without the notochord functioning as an organizer in an early vertebrate embryo, the development of the entire central nervous system of the individual is compromised. In the two other subphyla of the chordates, Cephalochordata and Urochordata, the notochord is either maintained in the adult (Cephalochordata such as Lancelets) or present in the larval form and lost during metamorphosis into the adult form (Urochordata such as Tunicates).

Another character that unites the chordates is the pharyngeal arches and pouches (sometimes inaccurately called gill arches and gill slits). It is not advantageous for terrestrial animals with lungs to retain the gills that develop from the pharyngeal arches and pouches in fish species. The successful embryological development of terrestrial vertebrates, however, requires the organizing presence of these early developmental structures. For example, among other things, the arches give rise to facial bones, parts of the inner and outer ears, and the cartilage of the larynx. The pouches give rise to the Eustachian tubes, and the thyroid and thymus glands. Proper interaction between the arches and pouches is required for normal embryological development. For example, a mutation on chromosome 22 in humans allows an improper association between these early developmental structures and results in the thymus failing to develop in the human embryo. The resulting syndrome, DiGeorge syndrome, causes recurrent infections, heart defects, and facial bone malformation.

### Conclusion

Whereas Haeckel's recapitulation theory was wrong in detail, the larger notion of recapitulation, as we now understand it, provides powerful support for Darwin's ideas of common ancestry and descent with modification. Because development is genetically controlled, small changes in developmental genes can have substantial implications for an individual's juvenile and adult form and function, and on its eventual ability to survive and produce offspring of its own.

In the next chapter, we will introduce and describe a new field of biology that shows, through the biology of genes, how an individual passes from a single-celled egg to a multicelled adult and how a fish journeys through evolutionary time to eventually arrive as a human. This is the field of evolutionary developmental biology, or Evo Devo.

## Thought Questions

1. Think of an explanation you once depended on to explain some phenomenon in nature (for example, why the grass is green), but later found out was incorrect. What created the incorrect explanation in the first place, and what needed to happen for you to finally get it right?
2. How did Haeckel's drawings of vertebrate embryos influence Darwin and his theory of evolution?
3. Explain why a human embryo might pass through a stage when it resembles a fish embryo but would not pass through a stage where it resembles an adult fish.

# Chapter 12

# *Evo Devo*

## How Evolution Constantly Remodels

Why are there so many different kinds of living things?

—Stephen Jay Gould, Harvard University, and any inquisitive five-year-old on a trip to the zoo

### From Noble Flies to Nobel Prizes

At the turn of the twentieth century, embryologist-turned-geneticist Thomas Hunt Morgan began work on understanding genetics using a tiny, neglected little insect called the fruit fly (*Drosophila melanogaster*; yes, the same one that's on your bananas right now!). The normal eye color for fruit flies is brilliant red. One day, however, in his "fly room" at Columbia University, Morgan looked into the vial of fruit flies that had been living for almost a year and discovered that one of the males had white eyes. This is the first documentation of the now famous white-eye mutation. Morgan and his students' relentless work on understanding this mutation led to the discovery that genes are in fact located on structures called chromosomes. In 1933, Morgan (but unfortunately, not a single fruit fly) was awarded the Nobel Prize in Physiology or Medicine for his work on fruit fly genetics. Following Morgan's lead, geneticists for decades have used the noble fruit fly for understanding everything from simple genetics to behavior and physiology, among other phenomena.

Up to the 1970s, embryologists thought that the genes that controlled body forms as apparently disparate as honeybees and hound dogs were as different as the organisms themselves. But geneticists Edward Lewis of Caltech, Christiane Nüsslein-Volhard of the Max Planck Institute for Developmental Biology, and Eric Wieschaus of Princeton University were not so convinced. In 1978, Lewis showed that a group of similar genes (a gene complex) called *bithorax* controlled the development of body segmentation in fruit fly larvae. Also while working on fruit flies in the 1970s, Nüsslein-Volhard and Wieschaus together began the daunting task of sorting out all the gene products essential to the development of the fruit fly embryo. They systematically eliminated one gene at a time and screened for developmental defects in the resulting mutant larvae. In one year of exhausting work in their lab at Heidelberg, Nüsslein-Volhard and Wieschaus established 26,978 unique lines of flies and tested each one for mutations. This technique of eliminating genes (called gene *knockout*) to identify and understand their functions is still used today, especially by biologists investigating the genes that control animal development.

The now famous *Heidelberg screens* revealed 139 developmental genes. The most revealing aspect of this work was that many of these genes could be grouped together because they affected the same developmental process. Nüsslein-Volhard and Wieschaus's work, coupled with that of Lewis, provided fascinating evidence for the genetic control of development and provided a launching pad for the new field of evolutionary developmental biology—the beginning of a revolution in our understanding of the genetic controls of development and their implications in evolution. As a result of their contributions to our understanding of the genetics of development, Nüsslein-Volhard, Wieschaus, and Lewis were jointly awarded the 1995 Nobel Prize in Physiology or Medicine.

We can thank millions of individual fruit flies for all of this knowledge. Even today, the fruit fly remains one of our major model organisms and is shedding light on the genetic mechanisms that drive, for example, cancer formation.

### GENES

Before launching too deeply into a discussion of the roles of genes in evolution, we first review the basics of what genes are and what they do. A gene is a specific segment of deoxyribonucleic acid (DNA—if not the blueprint, then the recipe of life) that indirectly provides the instructions for the specific proteins that a cell manufactures. The DNA does so from inside the nucleus (command center) of a cell by transcribing instructions using a similar molecule, ribonucleic acid (RNA). The specific sequence of the building blocks (adenine, guanine, thymine, and cytosine) that make up DNA determines the specific sequence of the building blocks of the RNA that leaves the nucleus. The RNA instructions are used to build specific proteins, many of which will perform specific tasks (by way of chemical reactions) for the cell, giving each cell a unique functional identity.

The particular genes turned on at a particular moment in the nucleus will determine which proteins are present in the cell and which chemical reactions take place. Often these proteins or their products can in turn also influence which genes are turned on or off in the nucleus and which additional proteins are manufactured. This elegant and incredibly complex cascade of signals is what ultimately determines how an individual organism develops as an embryo, the specific traits the individual ends up with, and how that organism ultimately functions as an adult. Francis Crick, the codiscoverer (with James Watson) of the structure of DNA, once described the process, DNA to RNA to protein to trait, as the central dogma of molecular biology. But, unlike religious or political dogmas, this one is based on science and is open to experimentation and correction if future evidence proves it wrong. In short, it describes the fundamental way that genetic information is expressed in cells.

### The Evolutionary Synthesis

The central dogma is a product of our understanding of the structure of DNA, but it was also born out of an earlier consensus by evolutionary scientists working on various angles of evolutionary biology, primarily in the period 1937–1947. During this time, naturalists, geneticists, paleontologists, and embryologists realized that they were all working separately to understand the mechanisms that underlay evolution. So they began sharing information and educating themselves in each other's fields. The consensus also included statistical approaches that served to quantify and predict the possible evolutionary outcomes of genetic changes in populations of species. Ernst Mayr has described the consensus as the *evolutionary synthesis* (also called the *modern synthesis*, the *great synthesis*, and the *new synthesis*). This advance in evolutionary thought is based on the fact that known genetic mechanisms can explain all known evolutionary phenomena, and it has both refined and broadened Darwin's original ideas. For example, while Darwin was wrong about the blending of traits from parents to offspring, the evolutionary synthesis strengthened his original idea of descent with modification by showing how changes in DNA can be passed from parent to offspring.

A subset of the evolutionary synthesis was the joining of single-gene genetics with the characteristics studied by naturalists, called *polygenic traits*, which vary across a spectrum of forms. We gave some examples of these traits (weight, speed, color) in our discussion on fitness in chapter 10. In its infancy, genetics was focused on single genes. For example, Gregor Mendel's experiments were on the single genes that determined plant height or seed shape, and even Morgan's work with the fruit fly began by following the inheritance patterns of single genes. But scientists began exploring more-complex genetics. While natural selection can act on the either/or traits that single genes influence, most evolutionary action focuses on traits that are controlled by more than one gene at a time—polygenic traits—like the patterning of body segments that are under the control of the *bithorax* gene complex, described earlier in this chapter.

### *Hox* Genes

As the new scientific discipline of evo devo grew out of the evolutionary synthesis, work on the genetic control of development branched into other model organisms. Figuring out the unique DNA sequence of a specific gene (through gene sequencing) and using an individual organism's entire genetic code to create another individual just like it (cloning) has also gotten easier, faster, and cheaper in recent years. It became more and more clear that the gene complexes controlling development in fruit flies also controlled development in other insects (e.g., bees), and in vertebrates, including fish and mammals. This versatility was even more evidence in support of Darwin's original ideas of the common ancestor and descent with modification. For example, after Lewis's discovery of the *bithorax* gene complex, biologists discovered an evolutionary connection that is predicted by descent with modification. They found that vertebrates, including humans, have a complex of genes called the *homeotic* complex, which is similar in DNA base sequence to *bithorax*. The homeotic complex has descended from the *bithorax* complex but has been modified by millions of years of mutation.

Homeotic (*Hox*) genes were then discovered to be master control genes. These genes share an identical 180-DNA base segment called the *homeobox*. In consequence, their protein products all share an identical region called the *homeodomain*, which allows specific proteins to bind to DNA and thus influence the activity of particular developmental genes. Mammals have thirteen *Hox* genes and fruit flies eight. We now know that *Hox* genes control the specific identity of each segment in a developing animal embryo by directing the fate of the cells in which the genes are active. Therefore, even the smallest of imperfections in a *Hox* gene can have catastrophic implications for later development. For example, the human disorder called DiGeorge syndrome mentioned earlier has its origin in a faulty gene in one of the *Hox* complexes.

*Hox* genes are so important in development that mutations in *Hox* genes can result in debilitating syndromes in individuals (eventually compromising their fitness), but more likely produce mutations that are lethal to the developing embryo. Because most mutations to *Hox* genes are lethal, these genes cannot evolve rapidly. As a result, the genes that control patterning in species from insects to fish to mammals are very similar, in both DNA sequence and function, to the genes originally discovered in fruit flies. In the jargon of evo devo–speak, the genes have been *conserved* through evolutionary time and connect different lineages to common ancestors. They therefore provide powerful evidence for descent with modification and evolution. To illustrate conservation, we next provide two very different examples, one between the fruit fly and a fellow insect, the honeybee (*Apis mellifera*), and one between the fruit fly and a complex mammal, the human.

Fruit flies and honeybees diverged from a common ancestor about 300 million years ago. As we know from our earlier discussion on speciation, ever since fruit fly and honeybee ancestors became isolated from each other, subsequent genetic changes in one lineage could not have been shared by the other. In 2000, the entire DNA sequence (the genome) of the fruit fly was published, and in 2006, the honeybee genome was published. Developmental biologist Peter Dearden and his colleagues at the University of Otego have since shown that the vast majority of developmental genes in both insect species have been conserved, even over 300 million years of evolution. The few differences in developmental genes between the two species—genes missing in the honeybee but present in the fruit fly—are probably genes that duplicated during the evolution of the fruit fly lineage. The duplicate genes then mutated and took on a new function. Dearden and his colleagues concluded that if a gene provided multiple functions during development, it was likely to be preserved in both insect lineages. They also hypothesized that genes that persist throughout the evolutionary history of a species are likely to evolve additional functions.

But why else should we be interested in the honeybee? The urgency of understanding the evolutionary history and developmental genetics of the honeybee is obvious once we consider the economic importance of honeybees. In the United States alone, honeybees pollinate billions of dollars in crops, from alfalfa to zucchini. The more we know about this important species, the more quickly and effectively we can act when populations of the species are faced with disease, reduction in population sizes, behavioral changes, and other challenges. For example, from 1947 to 2005, honeybee colonies in the Unites States declined by 40 percent, from 5.9 million to 2.4 million. Scientists have named the decline colony collapse disorder (CCD). Several factors have been implicated as partly responsible for CCD, such as viruses, bacteria, and pesticides, but a primary culprit has yet to be identified. The publication of the entire honeybee genome in 2006 has provided geneticists with the opportunity to test hypotheses for CCD at the molecular (gene) level.

If similarities in developmental genes between fruit flies and honeybees provide fascinating evidence for evolution, similarities between the developmental genetics of fruit flies and humans are truly mind-boggling, especially considering that the fruit fly and human lineages diverged nearly one billion years ago. Further, what we are learning from studying the evolution and development of fruit flies has provided remarkable insight into human development, the neural mechanisms of learning and memory, and perhaps most importantly, cancer.

As a result of fruit fly genetic research, we have discovered dozens of gene families that, if altered by mutations, send fruit flies along debilitating developmental pathways that are often lethal. When high school and college students study genetics and the probability of inheritance (Mendelian genetics), they often use the fruit fly, with its many famous mutations, as a model organism. Some of the classic mutations that are studied include *apterous* (flies without wings), *vestigial* (flies with shriveled, useless wings), *white* (flies with the white eye mutation made famous by T. H. Morgan), *ebony* (flies with dark-colored bodies), and *notch* (flies with a notch at the tip of each wing). All of these gene families found in the fruit fly are also found in one form or another in humans. They are essential to the control of development in fruit flies, but many are also active in adult human cells, regulating whether a cell will become cancerous. When particular genes in these gene families are altered, they can cause mistakes early during human embryological development. Alternatively, if the mutation in the gene is more subtle, the consequence of the mutation might not be apparent until adulthood when affected cells contribute to tumor formation. In other words, many developmental regulatory genes are now understood to have strong links to human cancers. The *notch* family of genes provides an excellent example of this phenomenon.

In fruit flies, the *notch* gene codes for a specific protein that ends up embedded in the cell membrane of the embryonic cells that give rise to wing tissue. The *notch* gene controls normal wing development through a series of complex cell-to-cell interactions and cascading molecular signals. An altered *notch* gene, however, produces an altered notch protein, which in turn produces an altered trait, the notched wing, in the adult fly. Even after nearly one billion years of evolution, the *notch* gene is still recognizable in humans, yet it has duplicated and evolved. Humans now have four of these *notch* genes—all different—and therefore four different notch proteins. These protein products of the human *notch* genes are from 11 percent to 64 percent identical to the protein coded for by the fruit fly *notch* gene.

In humans, *notch* gene mutations cause cancer. For example, one of these *notch* genes, *Notch 1*, is linked to a form of leukemia (cancer of the white blood cells). Another *notch* gene, *Notch 3*, is active in the smooth muscle development of vein and artery cells in adult humans. Hyperactivity of the gene, however, has been strongly linked to lung tumor formation and subsequent human lung carcinoma. It is remarkable that a wing-formation gene first discovered in the fruit fly has been found to cause lung cancer in humans. The presence of *notch* genes, like *Hox* genes, in both humans and fruit flies (and every species in between) is more evidence of evolution through common ancestry. Because of this conservation of genes throughout evolutionary time, we can use the noble fruit fly and clever experiments to figure out how to suppress the *notch* gene and avoid tumor formation in humans.

### THE ARRIVAL OF THE FITTEST

In a 2007 article in the *New York Times*, Scott Gilbert, a developmental biologist at Swarthmore College, explained that while classic evolutionary biology looks at the survival of the fittest, studies in evo devo show that it may actually be "the arrival of the fittest." Gilbert's contemporary, University of Wisconsin–Madison biologist Sean Carroll, explains in his book *Endless Forms Most Beautiful* that when we look at the fossil record and the amazing diversity in animal form that has come and gone over the millennia, an understandable (but incorrect) explanation might be that new genes must have evolved to create the new body we observe. In this view, the individuals with those new traits (and the apparently new genes that code for them) deemed "most fit" by the current environment were then selected to produce more offspring than individuals without the new traits and genes. This hypothesis was promoted by Edward Lewis after he discovered the *bithorax* genes. When Carroll and his students investigated this hypothesis, however, they found that instead of new genes appearing over time, the same genes were being used in different ways. In other words, they provided more evidence that evolution works by modifying preexisting genes to take on new roles required by changing environments.

Carroll first found that most of the species in the phylum Arthropoda (including spiders, centipedes, lobsters, and fruit flies) share all ten *Hox* genes with a distant cousin, the Onychophorans. Onychophorans are small organisms (most them are no bigger than a human hand) found in the tropics and subtropics. They resemble caterpillars and they have many jointed appendages like arthropods. However, the onychophoran and arthropod lineages diverged 500 million years ago, and the two groups are now very different from each other in form. How was it possible that these very different groups of animals use the same genes to control their embryonic development? Carroll found that the organisms were simply using the same *Hox* genes in different ways.

During the development of the embryos of these animals, a subset of the *Hox* genes are turned on (actively making their protein products) in some segments of the embryo and turned off in other segments. Which genes are on and which are off in each segment have changed over evolutionary time, resulting in different arrangements of appendages and segments, and therefore the appearance of very different body forms. For example, in arthropods, the segments in which the number eight and nine *Hox* genes are turned on, or *expressed*, have legs. When gene expression is turned off in these segments and replaced by the expression of the number seven *Hox* gene, another type of appendage, maxillipeds, mouthparts modified for handling food, develops instead.

We now understand that the implications of this versatility of when and where master control genes are expressed goes far beyond body segments and includes hearts, fingers, eyes, and other organs. These genes are so similar among distantly related animals that they can be transplanted from one organism to another without losing functionality. For example, Walter Gerhing and Kazuho Ikeo of the University of Basel showed that the master control gene that initiates eye development in the mouse, called *Pax 6*, can be inserted in the fruit fly genome to stimulate the fly's own series of instructions for eye development. *Pax 6* is now known to initiate eye development in organisms as different as sea squirts and humans. Even by itself, the presence and function of a gene like *Pax 6* among animals with eyes is evidence of common ancestry. That this gene is also found in animals without eyes is evidence that through descent with modification, genes can take on novel functions and result in a unique structure like the eye.

We also see evidence of this genetic predisposition in one of the most fascinating transitions discovered in the fossil record, evidence that answers the question, "How did fish learn to walk?" In other words, what evolutionary steps were required for the movement of vertebrates from aquatic habitats to habitats on land? The correct answer has been slow in coming. From over one hundred years of fossil discoveries, scientists had ideas about how the transition took place, but had not found fossils to support their hypotheses. But a recent fossil discovery and new evo devo research now give us a pretty good idea.

In 2004, on a northern Canadian island, paleontologist Edward Daeschler of the Academy of Natural Sciences in Philadelphia and his colleagues discovered the fossil remains of *Tiktaalik*, a vertebrate animal with limb bones, an elbow, and a wrist that were transitional between fish and *tetrapods* (four-limbed animals). What's more, this animal, which lived in shallow water habitats, had properties of both fishes and tetrapods: gills, scales, fins, lungs, a movable neck, and modified fins that could give it support either in water or on land. Not only was *Tiktaalik* a discovery of an intermediate between aquatic and terrestrial vertebrates, the finding was evidence that some aquatic vertebrates displayed potentially land-friendly structures long before the evolution of tetrapod vertebrates. For example, the hyomandibula is a bone in fish that functions as a necessary aid to gill-breathing, but in *Tiktaalik* the bone has lost this function and has begun taking on the function as an aid to hearing on land. The hyomandibula ultimately evolved to become the stapes bone in the inner ear of mammals. In other words, the limbs themselves and a future ear bone were present while tetrapod ancestors were still aquatic.

Soon after the discovery, scientists led by Zerina Johansen of the Natural History Museum in London showed a developmental connection between the digits (fingers) in living tetrapods and digit-like bones along the edges of the fin bones in living fish. They studied a *Hox* gene called *Hoxd13* that is essential for proper limb development in tetrapods (including humans) and fin development in fish. The scientists looked at *Hoxd13* expression in several living vertebrate species: salamanders (true tetrapods), lungfish (a fish that has lobed fins and can breathe air), zebrafish (a bony fish), and a *Polyodon* (a primitive paddlefish that appears in the fossil record 50 million years before dinosaurs and still swims in the Mississippi River today). The fact that *Hoxd13* expression was a requirement for inducing digit or radial bone development in all four groups is strong evidence that *Tiktaalik*, with its true digit bones, must have been under the influence of *Hoxd13* during its embryological development, just as living tetrapods are today.

We do not have DNA from the extinct *Tiktaalik* to confirm the presence of *Hoxd13*, but proteins, which are products of DNA, have been recovered from other long-extinct species and compared to animals alive today. For example, Mary Schweitzer, a paleontologist at North Carolina State University, and her colleagues recovered proteins from the 68 million-year-old leg bone of a *Tyrannosaurus rex* and compared them to similar proteins from several of today's animals. They discovered that the proteins were nearly identical to those of the modern chicken, which helps confirm that birds are descended from meat-eating dinosaurs.

It is now clear that the animal genome has had the potential for the remarkable biodiversity observed among animals today and throughout the fossil record for hundreds of millions of years; as Carroll argues in *Endless Forms*, "It is clear that genes per se were not 'drivers' of evolution. The genetic tool kit represents possibility—realization of its potential is ecologically driven." Therefore, the latest iteration of fitness for a population is simply waiting to be realized, waiting for the right combination of gene expression and selective pressure to set the stage for its arrival.

### Repeated Evolution

Carroll's statement also explains why we see *repeated evolution* of characteristics in many groups of organisms—groups that have been isolated from each other for thousands of generations. Repeated evolution occurs when at least two populations become isolated from each other, but later end up in similar environments with similar environmental pressures. As a result, the two (or more) populations follow similar adaptive paths. For example, dozens of populations of a small fish called the three-spined stickleback (*Gasterosteus aculeatus*) became isolated from their marine (saltwater) environments and from each other in individual lakes when the glaciers of the northern hemisphere retreated at the end of the last ice age. The freshwater lakes in which the fish became isolated were very similar in ecological characteristics (plant life and other animal life), and thus the isolated populations of sticklebacks evolved under similar environmental influences. Nature has therefore conducted an elegant experiment with several replicates and a huge sample size, nearly identical to what an evolutionary biologist would design if she were to do it herself.

One of Nature's experiments with Sticklebacks involves the protective armor that covers their bodies. Sticklebacks that live in marine environments have such protective armor, whereas individuals in many of the new freshwater stickleback populations have lost this protective armor. Evolutionary biologists think that when populations of the sticklebacks were first isolated from each other in separate lakes ten thousand years ago, a few individuals carried a mutant form of a gene that is involved in the formation of the fish's armor. Once in a less-dangerous environment, individual fish with the mutant form of the armor-producing gene were more successful than armored fish with the functional gene, probably because building the armor is energetically expensive and becomes a burden if it is unnecessary. The number of individuals in the population with the mutant form of the gene then increased dramatically with respect to the individuals with the functional gene. In turn, the proportion of the mutant gene to the functional gene changed in favor of the mutant gene (classic microevolution). This phenomenon was repeated independently several times in different lakes (repeated evolution).

Another stickleback trait that is advantageous in a marine environment is a pelvis with spines that poke through the fish's scales and make it difficult for predators to swallow the fish whole. As with the loss of armor, many populations of freshwater sticklebacks no longer have the spines or enough of a pelvis to produce them. However, unlike armor, the reduction of the pelvis and loss of the spines is not the result of a mutation; rather, the gene involved in the formation of the pelvis is expressed at different times and different places during embryological development. Michael Shapiro at Stanford University and other curious evo devo scientists have found in mice the same gene that controls pelvis and spine development in the stickleback and essentially turned it off. Sure enough, the mice were born with greatly reduced pelvis bones and hind limbs, just like the sticklebacks.

But groups need not be closely related for a behavior or trait to evolve repeatedly. For example, in species in which the female is promiscuous (mates with many different males), the males develop large testes and high sperm counts. In this *polyandrous* system, where many males mate with one female, the male that can deposit the most sperm is more likely to father the offspring of the female. Therefore, the large testes/high sperm count trait is likely to be passed on to the mating male's male offspring. We see this trait in species as different as yellow dung flies and Norway rats because both species have polyandrous mating systems. Even within the primate family, testis size

### RARE AS HEN'S TEETH

In 1980, Edward J. Kollar and C. Fisher of the University of Connecticut School of Dental Medicine, took epithelial (skin) tissue from the pharyngeal arches of chick embryos and grafted them onto embryonic mouse tissue. The details are too complex to repeat here, but several of the grafts generated teeth—hen's teeth. Later, in 2006, Mark Ferguson of the University of Manchester and his colleagues examined a mutant chicken and found a complete set of teeth that closely resembled crocodiles' teeth. Inspired by this discovery, they activated tooth genes in normal chick embryos and found that the embryos developed teeth. Because chickens have descended from toothed dinosaurs similar to *Archaeopteryx*, the tooth genes must have lain dormant in the chicken for approximately 80 million years.

A trait that has disappeared from a given species and then reappears in some descendants of that species is known as an *atavism*. Atavisms occasionally appear naturally, as when a horse is born with toes, a dolphin or whale is born with hind limbs, or a flatfish such as a fluke or a flounder hatches with camouflage coloration on the wrong side (that is, the underside, which is not visible to predators). Even though the adult tail disappeared from the ancestors of humans when the ape and monkey lineages diverged 25 million years ago, some humans are still born with intact tails and have them surgically removed. Human embryos have tails for a short time after the first month of embryological development, but the tail usually stops developing and is reduced to no more than the tailbone (coccyx) when humans are born.

Atavisms provide a pool of genes that remain in the genome but are not expressed. Occasionally, an organism born with an atavism may turn out to be more fit than the general population, perhaps because of a change in the environment. Atavisms may, therefore, perform an important evolutionary function.

Atavisms differ, incidentally, from *vestigial organs*. These are rudimentary organs that no longer function but descend from fully developed and functioning organs in the ancestral species. The human appendix is the classic example of a vestigial organ, but in fact it may have a function in maintaining the bacteria in the intestine. Better examples of vestigial organs include vestigial legs in certain primitive snakes and in dolphins and whales. The hair on your arm could also be considered vestigial, as could your toenails.

Atavisms and vestigial organs both provide strong evidence in favor of common descent. Embryos, as we have noted, recapitulate the embryonic development of ancestral species. Frequently, a new organ is built on the vestiges of an old organ, as when the gill arches provide the foundation for the facial bones, the ears, and so on (see main text). Sometimes, however, an organ remains purely vestigial or a gene is not expressed. It is very hard to explain the presence of vestigial organs and unexpressed but potentially functional genes without appealing to common descent.

fluctuates with mating system. Monogamous gorillas have tiny testes (and resulting sperm counts) relative to their body size, while promiscuous chimpanzee testes are comparatively large. Human males fall somewhere in between. The fact that human testes are relatively larger than gorilla testes is evidence that there may still be sperm competition in the human population. Fitness depends on an individual's ability to pass genetic information on to its offspring. Therefore, it is no surprise that general characters like armor (or a lack of it), spines, and a high sperm count have evolved repeatedly both among populations within a species and among different species.

## Convergent Evolution and the Evolution of Mimicry and Camouflage

Repeated evolution of similar structures and functions usually involves entire blocks of similar genes in populations of related species. However, environmental pressures can also influence the evolution of similar structures and functions in organisms that do not share a common ancestor, and these structures are built by entirely different sets of genes.

Take, for example, the torpedo-like shape of a dolphin (a mammal), an ichthyosaur (an extinct swimming reptile), and a penguin (a bird). Dolphins (and whales) evolved from terrestrial (land) mammals that adapted to a marine lifestyle. In a similar way, ichthyosaurs evolved from terrestrial reptiles. Penguins, being swimming birds, owe their ancestry to terrestrial dinosaurs. As they evolved from their different ancestors, all three groups were under the influence of the same laws of physics that operate in underwater habitats. While different, the genes controlling their shapes arrived at the same morphological conclusion: the shape of a torpedo. This shape, shared by all three (as well as numerous other marine species, like the shark and the squid), allowed for efficient movement through water at high speeds.

Evolutionary biologists call the phenomenon of unrelated species sharing analogous characteristics *convergent evolution*. Examples of convergent evolution are found across many unrelated groups of organisms: the wings of bats, birds, and insects all serve the same function of flight; an extra flap of skin found between the fore and hind legs of the North American flying squirrel (a mammal) and the Australian flying phalanger (a marsupial) allows both to glide from tree to tree; and the spines on the branches of locust trees and the stems of rose bushes help both plants avoid getting eaten.

Perhaps even more fascinating is the evolution of mimicry to avoid predation. The discovery of mimicry in 1862 in Amazonian butterflies by Henry Walter Bates was perhaps the first solid proof of natural selection. Bates found that one species of palatable (nonpoisonous) butterfly had evolved the same body coloration and patterning as that of a noxious (poisonous) species. Bates discovered that the body patterning of the noxious species varied among populations across Amazonia—and so did the palatable species. Natural predators learned to avoid both species, even though the palatable species would have been fine to eat. Mimicry is found across animal taxa. For example, the rear end of the hawk moth caterpillar resembles the head of a tree snake. Birds that have learned to avoid snakes also avoid the hawk moth caterpillar and miss out on a nutritious snack.

The evolution of camouflage (cryptic coloration) has also allowed organisms to avoid being eaten by escaping detection by potential predators. This evolutionary strategy is relatively common, and examples range from the simple earth tones of desert animals to the complex patterns on the surface of fish and reptiles that have evolved to blend into substrates upon which they rest. Camouflage in many cases is also a form of mimicry. For example, the wing colors and shapes of leafhopper and plant hopper insects resemble the small leaves of the plants on which they feed, and some even resemble the thorns of the stems on which they travel. The stone plants of southern Africa are simply two succulent leaves fused together and avoid getting eaten by looking like the stones on the intermittent stream bottoms where they grow. One of the most amazing examples of the evolution of camouflage/mimicry comes from the leafy sea dragon, a species of sea horse. This fish has lobes of skin that grow out in all directions and result in the appearance of the floating seaweed among which it swims.

All of these traits have their combined origin in the genes that build them and the environmental pressures on the organisms. Through our evolving understanding of *Hox* genes, we now know that small genetic changes can induce changes toward such complexity that an organism's physical traits can resemble something else in nature almost precisely. The evolution of complexity may, at times, seem nearly impossible. Yet complexity is possible, and new observations and experiments to test it continue to support its evolution.

## The Evolution of Complexity

How do complex genes, complex structures and functions, and even more complex genetic signaling pathways evolve, when, if one part is missing, the whole system can break down? Darwin himself worried that if evolution did not progress in a stepwise process, his theory would have little to stand on. In *Origin* he wrote, "If it could be demonstrated that any complex organ existed which could not possibly have been formed by numerous, successive, slight modifications, my theory would absolutely break down." As we discussed in chapter 8, some detractors of evolutionary theory propose the idea of irreducible complexity: that some biological structures, such as the bacterial flagellum, are so complex that they cannot have evolved gradually. Refuting this claim, evolutionary biologist Joseph Thornton of the University of

Oregon and his colleagues showed in a study published in *Science* in 2006 that complex molecular systems evolve through a clever yet normal selective process. They call this process *molecular exploitation.*

Thornton's group studied the hormone aldosterone, which regulates blood pressure in humans and is found in other vertebrate tetrapods. For aldosterone to induce the necessary response in a cell, it must bind to a very specific receptor (a binding partner) on the cell's surface; no other receptor found in tetrapods works. The scientists wondered if aldosterone would communicate with an ancient receptor, the same receptor from which the tetrapod receptor evolved. To test this hypothesis, Thornton and his colleagues chose lampreys and hagfish, both jawless fishes that are alive today but whose common ancestor diverged from the ancestor of the tetrapods over 450 million years ago. Lampreys and hagfish do not use aldosterone, but still have an ancient form of the aldosterone receptor. Thornton and his colleagues took this living yet ancient receptor and found that it could in fact be stimulated by the aldosterone from tetrapods. They then reconstructed the jawless fish receptor (through a process called *gene resurrection*) that would have existed 450 million years ago and tested it with aldosterone. This ancient receptor also could communicate with aldosterone. But aldosterone did not exist half a billion years ago. The ability for tetrapods to even manufacture aldosterone evolved only recently.

What makes Thornton and his group's work so exciting is that they have shown that when a novel stimulus molecule (in this case, aldosterone) shows up, perhaps by way of a slight change in an existing stimulus molecule, it can exploit an ancient (but still existing) receptor molecule and recruit it as a binding partner. The ancient stimulus molecule was already present because it had been selected, long ago, for an entirely different function. This assertion is also supported by the fact that most biological molecules are capable of multiple functions anyway.

Therefore, the complex and specific aldosterone-receptor interaction and resulting cascade of molecular signals we observe today in tetrapod cells did not have to evolve at once, which is highly improbable, nor did they need to be magically put in place by a designer.

Thornton and his colleagues close their breakthrough paper by explaining that old molecules with unique ancestral functions can combine with new molecules and create a tightly integrated, complex system. They argue that this molecular exploitation "will be a predominant theme in evolution, one that may provide a general explanation for how the molecular interactions critical for life's complexity emerged in Darwinian fashion."

## Conclusion

The new discipline of evolutionary developmental biology provides clear, empirically based answers, not only for why there are so many different kinds of living things, but also how these things come to be, through genetic control during embryological development. Evo devo is shedding light—floodlights even—on how the smallest and most intricate of life's processes evolve. It is also changing the ways we arrange past and present species and their evolutionary relationships in the tree of life. In the next chapter, we will show how molecular biology has forced paleontologists and evolutionary biologists to cut down their old "trees" (called phylogenetic trees) and build new ones.

### Thought Questions

1. Think of how the fruit fly has been used for more than a century as a model organism for understanding biological phenomena. What characteristics make for a good model organism? What future discoveries might be possible as a result of studying model organisms?
2. Why are so many genes conserved through evolutionary time?
3. Explain why regulatory genes are more likely to be conserved than other genes. What genes are least likely to be conserved?

CHAPTER 13

# *Phylogenetics*

We are all but recent leaves on the same old tree of life and if this life has adapted itself to new functions and conditions, it uses the same old basic principles over and over again. There is no real difference between the grass and the man who mows it.

—Albert Szent-Györgyi, Nobel Laureate in
Physiology or Medicine, 1937

SOMETHING HAPPENED over the last 3.8 billion years to create the biodiversity found on earth today. The job of evolutionary biologists and paleontologists is to ascertain what happened, how it happened, and why it happened. In other words, their challenge is to reconstruct the patterns and the history of evolution, and one of the most illustrative products to come out of this quest is the "tree of life," a model of the evolutionary relationships among past and present organisms. The first trees were constructed not by Charles Darwin, but by Ernst Haeckel, who coined the phrase tree of life (see chapter 11), to show common descent within groups of organisms by looking at *homologous* structures—structures possessed by two or more species, like the wing of a bat and the foreleg of a horse, that have been derived, with or without modification, from a common ancestor.

## TREES OF LIFE

At first blush, trees of life, or *phylogenetic trees*, look like simple constructs. And indeed, evolutionary biologists use the guiding principles of maximum parsimony and maximum likelihood when they build phylogenetic trees. The *principle of parsimony*, sometimes called *Ockham's razor*, was first enunciated in the fourteenth century by William of Ockham. Applied to biology, the principle of parsimony argues that the arrangement of species in a phylogenetic tree involving the fewest unsupported assumptions is the most probable arrangement—the simplest explanation is the most likely, or at least the one deserving of our first attention. Because these trees are based on the best available data at the time they are constructed, they are hypotheses for how past and present biodiversity may be related. We continually test these hypotheses with new data and rebuild the trees whenever necessary.

The selection mechanisms underlying the similarities and differences among past and present species are poorly understood. As a result, it is at best incomplete to use only the fossil record and the structures of living organisms to reconstruct evolutionary history. Therefore, phylogenetic trees will surely change as we gather and analyze more data, especially molecular (DNA) data. These changes in evolutionary hypotheses are misunderstood by creationists, who point to the changes as evidence of flaws in evolutionary theory. Yet the progress of science depends on its characteristic of dynamic flexibility as we gather and analyze new data and test new hypotheses. In this section, we will discuss three examples of phylogenetic trees, how they were originally constructed, and how they have changed as new data roll in and are analyzed. First, the big tree, and then two of its branches (twigs, more like).

*The Big Tree*

The first highly organized life forms, single-celled organisms, are fossilized in rock strata that are 3.5 billion years old. Many of these bacteria, which we call *prokaryotes*, diversified and began using photosynthesis as their primary mechanism for harvesting energy for metabolism. Photosynthesis slowly replaced the oxygen-poor atmosphere of primitive Earth with one rich in oxygen. This replacement made the atmosphere increasingly toxic for organisms not adapted to living in the presence of oxygen, but it paved the way for the evolution of *aerobic respiration* (oxygen breathing) in other organisms. In order to perform the complex reactions of aerobic respiration efficiently, single-celled organisms had to develop additional structures and membranes to help organize and manage the new complexity. As the power of microscopy and the technology for studying molecules increased in the last hundred years, we have been able to make sense of the details of aerobic respiration.

With only primitive technology, however, the first evolutionary relationships among living and fossil organisms were worked out using only the limited number of characteristics, or *characters*, that could be detected. For example, before Anton van Leeuwenhoek brought the microscope to the attention of biologists in the mid-1600s, only characters visible to the unaided eye could be used to group organisms. Fossil organisms could provide only the characters that had been preserved (for plants, mostly woody material; for animals, mostly bones). Fossil organisms displaying similar characters were linked to common ancestors from whom those characters seemed to have been derived. Accordingly, scientists grouped modern animals and plants, with the same (or similar) characters as the fossils and linked them to the same common ancestors.

Under this scheme, scientists made mistakes, like grouping the crinoid fossil (an animal) with plants because it resembled a flower on the end of a long stalk. Under strong influence from Haeckel and other embryologists, comparisons were extended to characters found during embryological development. Still, scientists during the time of Haeckel gave little consideration to the physiology or ecology of organisms. This omission led scientists to confuse animal characters that had been simplified over evolutionary time with characters that were truly primitive. For example, some organisms are *sessile*, wherein the adult form is connected permanently to a substrate, like a rock, while other organisms are mobile and move about freely. Organisms can also be *parasitic*, wherein the adult form lives off another organism, the host, while other organisms are *free living*, or not dependent on living on or in another organism. Scientists incorrectly assumed that the sessile and parasitic lifestyles were primitive characters and that the characters of being mobile and free living were more recently evolved.

Between the time of van Leeuwenhoek and Haeckel, Carolus Linneaus, a Swedish doctor of medicine, developed a hierarchical system of classifying living organisms that he called *systema naturae*. This system, first published in 1735, included a nested design of most inclusive (general) to least inclusive (specific) groups. We still use a modified version of his original seven levels (kingdom, phylum, class, order, family, genus, species) in our modern system of classification (also called *taxonomy*). Linneaus, however, was not interested in fossils and rejected evolution. In fact, Linneaus's primary motivation for classifying living organisms was, in his words, "for the greater glory of God."

Scientists used the presence or absence of cellular pigments and/or specific cellular modes of locomotion to classify single-celled organisms and multi-celled molds and mushrooms. For example, some microscopic organisms use a whip-like flagellum and its rotary motion to propel themselves through their aquatic habitat. Other microscopic organisms are covered with hundreds of tiny hairs called *cilia* that beat rapidly back and forth, allowing the organisms to spin and tumble quickly through the water. Others move by causing their fluid insides to flow in a specific direction in what is called *amoeboid movement*. In this way, scientists used characters—or a lack thereof—to decide where organisms should be placed on the tree of life. They gave little attention to the comparison of cellular components and their structures and functions, mostly because of the limitations of microscopy. As a result, throughout the first half of the twentieth century, most scientists agreed upon a two-kingdom system of classification—the Plantae and Animalia—and a resulting phylogenetic tree with two major branches and many smaller twigs (figure 11). Bacteria, because they were obviously not animals, were grouped with the plants; basically, what was not animal was plant and vice versa. Yet, the discovery of bacteria eventually led to an even more basic grouping, that

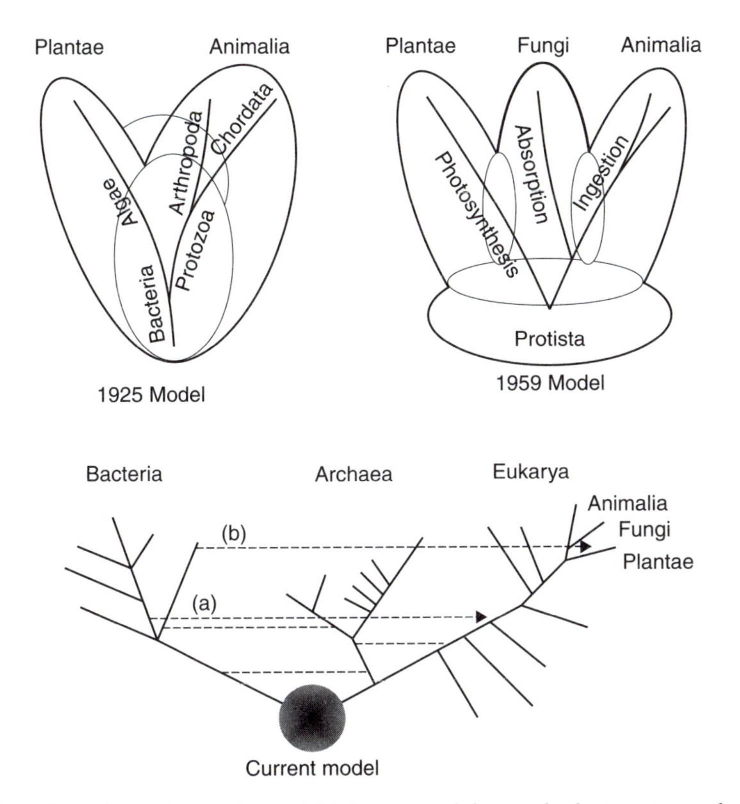


11. Three hypotheses for the "tree of life" presented during the last century. The upper left tree (1925) shows the two-kingdom (plant and animal) system of living organisms, with bacteria (prokaryotes) as the most primitive organisms. The upper right tree (1959) proposes a four-kingdom system, but with ecological roles as grouping characters. In 1977, Carl Woese and George Fox proposed the three-domain system (bottom) that included a newly discovered group, the Archaea (no nucleus), which shares a more recent common ancestor with Eukarya (nucleus) than either share with the Bacteria (no nucleus). The top two trees are based on ecological roles, as well as various combinations of embryological, physiological, and cellular structures that were visible with the microscopic technology available at the time. The bottom tree is based on similarities of molecular (genetic) material. The common ancestor to all three (gray circle) is unknown and may never be known because true cells (membrane-bound, independently reproducing collections of interdependent molecules) did not exist early on, and lateral gene transfer, the transfer of genetic material among groups (indicated by dotted lines), was common. Many symbiotic relationships were also established; in particular, symbiotic relationships with the Bacteria gave rise to (a) mitochondria and (b) chloroplasts in the Eukarya.

of prokaryote (single-celled organisms without a nucleus, all of which were bacteria) and eukaryote (single and multi-celled organisms with a nucleus).

A new tree of life proposed in 1959 by R. H. Whittaker of Brooklyn College included data on the structures and functions of the cellular components. This tree had three major branches (with accompanying twigs) that now grouped the molds and mushrooms into their own kingdom—Fungi. The three branches extended from a large trunk, the Protista, which included all single-celled organisms. The basic grouping of prokaryote (bacteria) and eukaryote (everything else) was still employed. But, by the middle of the 1960s, DNA and RNA base-sequencing technology was rapidly advancing, and scientists began to question the Whittaker tree-of-life hypothesis. The advance in technology was coupled with the discovery of the methanogenic bacteria. These single-celled organisms are anaerobic (do not use oxygen for metabolism) and consume carbon dioxide while producing methane gas; they are accordingly called *methanogens*. All three groups, the prokaryotes, eukaryotes, and now methanogens, use intracellular structures called *ribosomes* to manufacture proteins. Ribosomal RNA (rRNA) is composed of specific sequences of the bases adenine, uracil, guanine, and cytosine. The ribosomes in each group display very different base sequences from the rRNA in each other group. Comparison of other molecular characters further supported the definition of a third grouping. As a result of these new data, Carl Woese of the University of Illinois and his former research associate George Fox proposed in 1977 the domain system of classification, which now included the Archaea (where the methanogens were placed) as a new domain grouping. Additional data that have been collected since 1977 still support this three-domain system. As more and more molecular data are analyzed, however, the three-trunked tree that at first appeared sharp and clear is now beginning to blur because evolution at the single-cell level does not work in as predictable and straightforward a way as previously thought. Our continued analyses of molecular data among the Archaea and Bacteria have led some prominent biologists to consider the term prokaryote as misleading. For example, the evolutionary biologist Norman Pace of the University of Colorado argues that the prokaryote category is misleading in terms of evolutionary relatedness because it unites two groups, the bacteria and the archaea (see below), that do not share a most recent common ancestor. Pace suggests dropping the term altogether.

*Reticulate Evolution*

When microbiologists analyze the DNA from members in one domain, they find similarities with members in the other domains (e.g., plants with animals with archaea). The simplest explanation for this similarity has been that pieces of DNA and sometimes entire genes were transferred between domains. This process is now known as *reticulate evolution*. The initial reaction by many scientists to these new data was that the sky is falling. However, as Woese argued in a 2000 paper published in the *Proceedings of the National Academy of Sciences*, "When the scientific sky falls . . . the light dawns."

Microbiologists and biochemists have known for decades that single-celled organisms, which rarely transfer DNA by way of sexual reproduction, can receive foreign DNA by way of three major processes: *transformation*, *transduction*, and *conjugation*. In transformation, a bacterium can pick up pieces of DNA directly from its environment. Bacteriologist Frederick Griffith first demonstrated this process in 1928, when he showed that a nonvirulent form of *Streptococcus pneumoniae* could pick up pieces of DNA from a heat-killed virulent strain of *S. pneumoniae* and transform it into a virulent strain. Transduction involves infection by bacteriophages (phages for short), or viruses that infect bacteria. A phage that infects one bacterium can accidentally bring with it genes from that bacterium when it reproduces and breaks out of the cell. These genes can be incorporated into the genome of an unrelated bacterium if the phage then infects it. Conjugation, while asymmetrical, is the most nearly sexual transfer of DNA between bacteria. In this process, a physical connection is established between two bacterial cells, and genetic material is transferred directly from one cell to the other.

These processes are collectively called *horizontal gene flow* or *lateral transfer* and are shown as horizontal lines in figure 11. Scientists think that variations of these processes occurred during the early evolution of the three domains, blurring the trunks as we move back in evolutionary time toward the last universal common ancestor that unites all three. Scientists think that most of the horizontal gene flow must have occurred long before fully integrated cells existed—long before the incorporation of a new gene into the genome of a complex cell could cause serious and possibly lethal problems.

*Endosymbiosis*

It is clear that there must have been movement of genetic material from one domain to another. Yet we now have fascinating evidence that even more drastic events must have occurred: the transfer of essentially whole organisms into other organisms. Perhaps one kind of organism was accustomed to eating another or parasitizing another, but eventually two organisms developed a symbiotic relationship, in which one organism became a part of another. The development of symbiotic relationships (in which two different organisms collaborate and form a relationship of reciprocal helpfulness) between organisms eventually led to unions of very different organisms—prokaryotes being engulfed by eukaryotes, for example. The union of two very different organisms was discovered by the biologist Lynn Margulis while a young faculty member at Boston University. This union is now called *endosymbiosis*; it explains the presence of mitochondria and chloroplasts in eukaryotic cells. These once-autonomous *organelles* now function to transform energy within the cell. Mitochondria harvest the energy stored in biological molecules like carbohydrates and, by means of cellular respiration, make it available for the cell to power other vital functions. Chloroplasts take in light energy and transform it to chemical energy in the form of carbohydrates.

Mitochondria and chloroplasts are unique among all other cellular structures—they are like bacteria, in fact. Among many similarities, they have their own DNA and make many of their own proteins. Their DNA is also structured as a circular molecule like bacteria's (unlike the DNA in the eukaryote nucleus, which is in strands and organized into chromosomes) and has similar base sequences to bacterial DNA. Mitochondria and chloroplasts themselves even resemble bacteria in their size and physical structure. Scientists think that aerobic bacteria that were capable of cellular respiration and required oxygen (similar to today's purple bacteria, a group that includes *Escherichia coli*) developed symbiotic relationships with early anaerobic forms of the eukaryotes (called proto-eukaryotes) that were poisoned by oxygen. The anaerobic bacteria would have protected the aerobic *symbionts* and provided them with ingested food, while the aerobic symbionts would have protected the anaerobe from the toxic effects of oxygen. Similarly, photosynthetic bacteria (similar to today's cyanobacteria) developed symbiotic relationships with aerobic eukaryotic cells already housing mitochondria. These relationships provided the eukaryotes with the ability to harvest light energy, increasing their fitness relative to eukaryotes without the photosynthetic bacteria symbionts. This increased fitness led to the evolution of algae, the ancestors to plants. The eukaryotes provided protection and possibly mobility for the photosynthetic symbionts.

As discussed above, phylogenetic trees are hypotheses and guaranteed to change as we collect new information. We have shown how the general tree-of-life model has changed drastically over the last century, but smaller portions of the tree are changing constantly as we launch new experiments and ask more questions.

*Reshaping the Animal Tree*

In their chapter on animal diversity in the eighth edition of *Biology*, published in December of 2007, Neil Campbell and Jane Reece present two phylogenetic tree hypotheses to explain how the major phyla of the animal kingdom are related. Here the authors remind students that "the uncertainty inherent in these diagrams is a healthy reminder that science is a process of inquiry and as such is dynamic." One tree is based more on morphological (body system form and function) and developmental comparisons and had been the leading hypothesis up to 1988 (and still had proponents through papers published in 2001). 1988 was the first year in which molecular biologists published the first phylogenetic trees for animals based on ribosomal RNA sequence data. The other tree is based mainly on the new molecular data that have been pouring in in recent years. Even today, as you read these words, new data will support the latest tree, require that we modify a twig or branch of the tree, or even suggest significant changes in its structure.

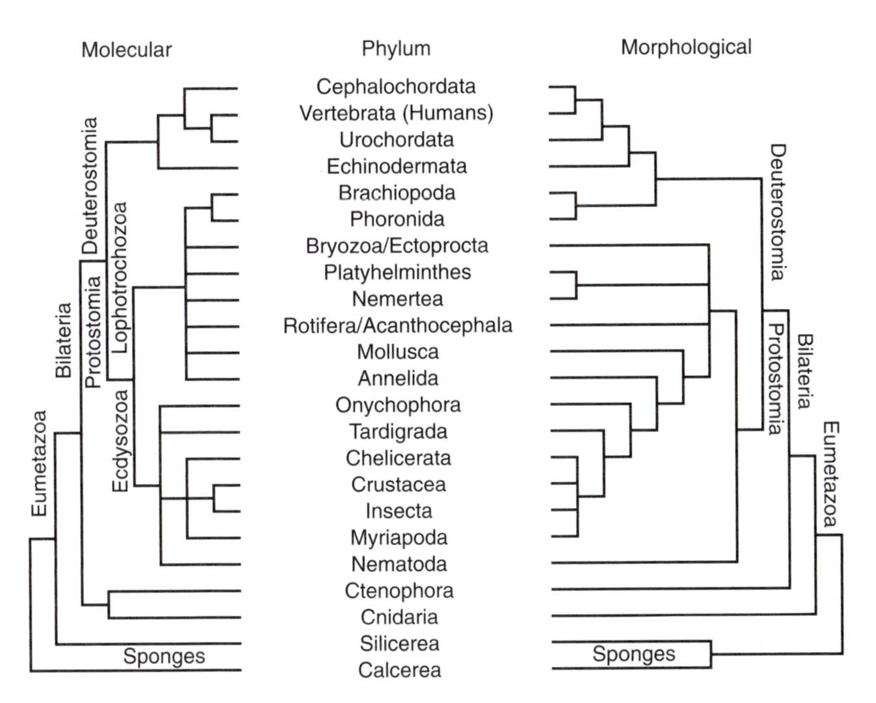


12. Two hypotheses for the animal phylogeny, one based on molecular data (left) and one based on morphological data (right). The list of animal groups in the middle is a selection of the more well-known animal phyla.

Therefore, the new molecular data can be used to test the morphological tree. Our sketches of both the morphological tree and the latest (2008) proposed molecular tree are illustrated in figure 12. To simplify the tree and highlight major similarities and differences, we include most of the more well-known animal groups but for simplicity exclude many others, which do not change the fundamental form of the tree.

Both trees—morphological and molecular—agree on several relationships as many of the morphological hypotheses pass the molecular tests:

1. All animals share a common ancestor, the uniting characters being that they are multi-cellular, eukaryotic (cells with a nucleus), and *heterotrophic* (cannot make their own food).
2. Sponges branch off from the base of both trees (thus they are *basal* animals).
3. All animals that are not sponges are linked to a common ancestor that acquired true tissues (the Eumetazoa).
4. Most animal phyla have bilateral symmetry (in which an organism can be divided into roughly mirror image halves, a left and a right) and are linked to a common bilateral ancestor (the Bilateria).

5. All members of the subphylum Vertebrata (animals with backbones), as well as many other phyla, are grouped into the Deuterostomia (second mouth). That is, the second opening to form during embryological development becomes the mouth of the adult, whereas the first opening becomes the anus.

The morphological and molecular trees differ, however, on some major arrangements. Here are two examples:

1. In the molecular tree, the Protostomia (invertebrate animals in which the first opening to form during embryological development becomes the mouth) have been divided into two major groups, the Ecdysozoa (animals that secrete an external skeleton that they occasionally shed, or molt, and replace as the animal grows) and the Lophotrochozoa (animals that have either a structure called the lophophore or that go through a larval stage called the trochophore). The animals with lophophores and trochophores are more genetically similar to each other than either is to the ecdysozoans.
2. In the morphological tree, the cephalochordates (animals with an internal supporting rod, the notocord, and a "head") are placed next to the vertebrates and linked by a common ancestor. The tunicates (animals whose larval form has a notocord and "head" but whose adult form does not) branch off earlier. In the molecular tree, the opposite is true; the tunicates are shown sharing a more recent common ancestor with the vertebrates instead of the cephalochordates.

As new data are presented at scientific meetings and in peer-reviewed papers, lively discussions are ignited as scientists agree and disagree. Eventually, consensus emerges. Scientists arrive at agreements on the shapes of phylogenetic trees because grouping animals using morphological characters is mostly subjective ("It is my opinion that these two fossils have similarly shaped jaw bones") while interpreting molecular data (DNA or RNA base sequences) is mostly objective ("The sequences for the wing development gene from these two insects is 99.9 percent the same. They must share a common ancestor around 20 million years ago"). With molecular data, we can be more certain that the evolutionary relationships we propose are truly the relationships that exist. All of this reiterates how dynamic and exciting the scientific arena of evolutionary biology is.

To this point we have shown how trees of life change through time. Now we will illustrate how, using a group of island lizards, phylogenetic relationships of species even in the same genus become increasingly clear as we add more and more data. And we will show how additional data and even different lines of evidence often validate earlier hypotheses.

*Building a Tree from Scratch*

In the early 1980s, evolutionary ecologist Roger Thorpe of the University of Wales, Bangor, began work on the ecology and evolution of lizards on the Canary Islands. He focused his efforts on the western Canary Island lizard (*Gallotia galloti*). This lizard species was of particular evolutionary interest because large populations of the species can be found on all of the four western Canary Islands: Tenerife, La Palma, El Hierro, and La Gomera. Thorpe and his colleagues wondered if the common ancestor to these populations was shared with two similar species, *G. stehlini* and *G. atlantica*, that populate the two eastern Canary Islands, Gran Canaria and Fuerteventura. The islands upon which the populations can be found are unique because they are volcanic in origin and are a product of the South Atlas fault line, which extends west off the coast of northwest Africa. Beginning 15.7 million years ago and ending 1.2 million years ago, each island erupted independently of the others in an east-to-west direction. The islands have never been connected by land bridges. Plants and animals from mainland populations have slowly colonized the islands since the islands emerged from the sea. Since colonizing the islands (most likely on rafts of debris carried seaward after storms), the western Canary Island lizards have evolved substantial variation both within the populations on each island and between populations on separate islands.

Thorpe and his students pieced together the evolutionary relationships of the populations of the lizard using data from various sources. Figure 13 shows Canary Island geography and the geographical distance, geological age of each

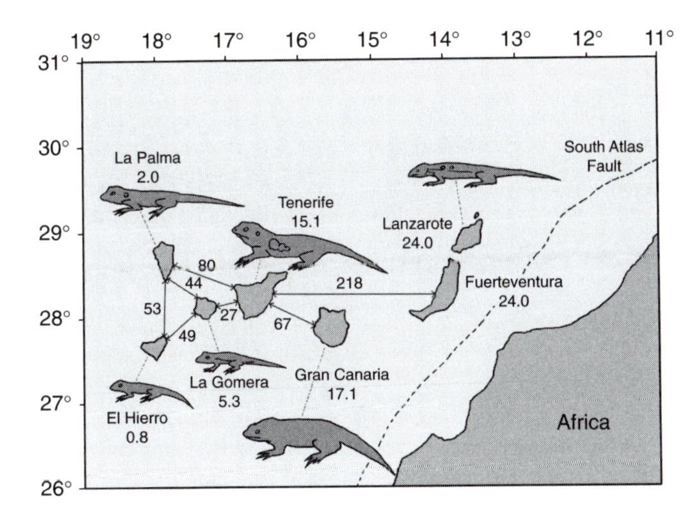


13. Canary Island geography showing the geographical distance (in kilometers), the geological age (in millions of years) of each island, and the relative body size of each *Gallotia* species. The ticks show latitude and longitude in degrees. Drawing by Alyssa Wiener.

island, and the relative body size of each *Gallotia* species. Figure 14 shows four phylogenetic tree hypotheses based on the independent analyses of various types of data. If we look at each source of data in a stepwise manner, we can see how each successive, independent hypothesis provides support for the one before, but also offers some contradictions. In all, each builds on the other to provide a best guess as to the evolutionary relationships of the lizard populations.

First, populations of the lizard could colonize the islands from the mainland of Africa, or from an older island, but only after each island emerged from the sea. More likely, though, lizards colonized islands near to the mainland before colonizing islands far from the mainland. Therefore, when we use only island age and distance from the next-nearest island to form relationships among the lizard populations, we come up with a phylogenetic prediction like the one illustrated in phylogenetic tree (a) in figure 14. In this hypothesis, we assume that the lizards colonized the islands in an east-to-west pattern; the

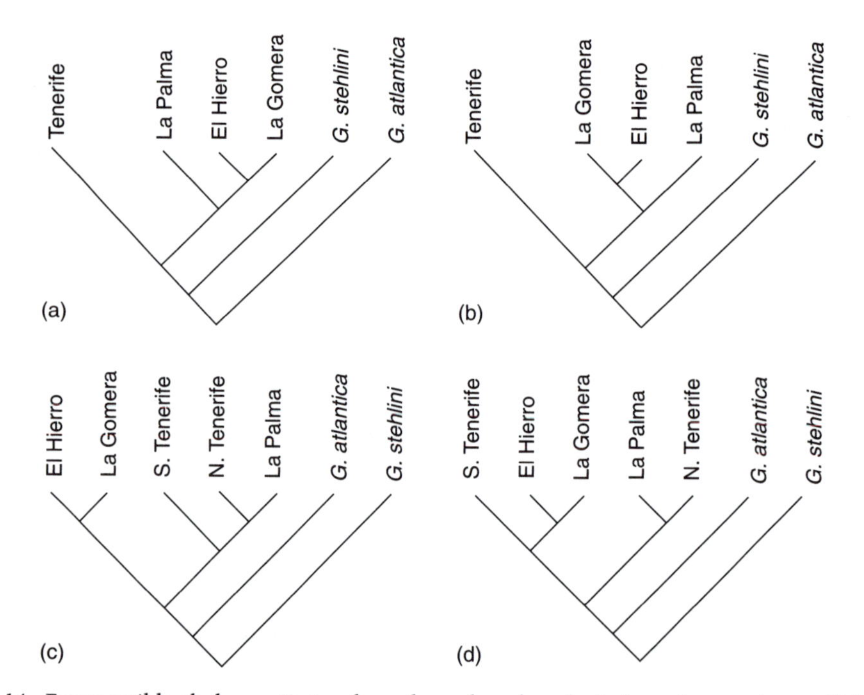


14. Four possible phylogenetic tree hypotheses based on the independent analyses of (a) geographical distance and geological age, (b) body size morphology, (c) cytochrome B sequence similarities, and (d) Thorpe's analysis of several sources of molecular data. (Redrawn from Thorpe, Roger, et al. 1994. "DNA Evolution and Colonization Sequence of Island Lizards in Relation to Geological History: Mtdna Rflp, Cytochrome B, Cytochrome Oxidase, 12s rRNA Sequence, and Nuclear Rapid Analysis." *Evolution* 48: 230–240.

only question is whether La Palma or El Hierro was colonized first from La Gomera. In our hypothetical tree, we assume that it is more likely that La Palma was colonized first because it is larger and older than El Hierro.

Second, in tree (b) of figure 14, we have arranged the lizard populations using only the morphological character of body size. Here we have assumed that the medium-sized La Palma population is directly descended from the larger Tenerife population and that the smaller La Gomera and El Hierro populations are descended from La Palma lizards. This arrangement both supports the hypothesis we present in (a) and also contradicts it.

Third, in tree (c) of figure 14, we have recreated a figure from one of Thorpe's papers published in 1994. This hypothesis is based on the DNA sequence of a gene called *Cytochrome B* found in the mitochondria of all eukaryotes and aerobic prokaryotes. The analysis of these molecular data reveals a detail not available from the morphological and geological data. Here we see that the Tenerife population is actually two genetically distinct populations—north and south types—with very little gene flow between them. The phylogenetic tree hypothesis that results shows that the La Palma population may have been colonized by ancestors of the North Tenerife population. The tree also suggests that ancestors of the South Tenerife population colonized La Gomera, and then individuals from La Gomera colonized El Hierro. The big surprise from analyzing the *Cytochrome B* data is that the most recent common ancestor to the western Canary Island lizard may in fact be *G. atlantica* from Fuerteventura instead of the geographically nearer *G. stehlini* from Gran Canaria.

Finally, when Thorpe and his colleagues test one line of evidence, the *Cytochrome B* hypothesis, with more molecular data, the hypothesis holds up, as does the hypothesis that the ancestors of *G. gallotia* most likely came from Fuerteventura and perhaps from Europe instead of northwest Africa. Further analysis of ocean currents supports this last point.

## Conclusion

Phylogenetic trees are hypotheses. They are our best guesses based on our analyses of the most recent and best available data. They are incomplete stories of evolutionary relationships, guaranteed only that they will change as we improve our molecular technologies and computational power and analyze more data. The more scientists present their ideas and discuss their assumptions, the closer we come to consensus. In a sense, like populations through time, phylogenetic tree hypotheses also evolve by way of descent with modification.

In the next chapter, we will discuss why nature appears to be designed and describe examples of both well-designed characteristics of organisms and characteristics whose "designs" present cumbersome challenges to their bearers.

THOUGHT QUESTIONS

1. Think of a critical decision you made at a point in your life that you based on the data available at the time. What critical data were you missing that may have changed your decision and how would your decision have been different?
2. Why do you think scientists and nonscientists alike are so interested in understanding the evolutionary relationships among both past and present organisms?

CHAPTER 14

# *Design by Committee*

## THE TWISTS, TURNS, AND FLIPS OF HUMAN ANATOMY

> In a perfectly designed world—one with no history—we would not have to suffer everything from hemorrhoids to cancer.
>
> —Neil Shubin, University of Chicago

IN HIS BOOK *Your Inner Fish*, Neil Shubin, a biologist at the University of Chicago, argues that "our humanity comes at a cost." What Shubin is referring to are our numerous fumbled, frail, faulty, and often downright useless structures and functions, many of which have their origin in our fish ancestors. As our evolutionary tree grew and added branches, fish were the first group to evolve jaws, paired appendages, paired sensory organs, a backbone, and even a head. We humans can be thankful for most of these developments. Yet with our kinship to fish, we also inherited some characteristics that are not well-suited to mammals. As a result, humans have had to evolve "designs" that at first blush seem not only clever but also purposeful. To the untrained eye, the human knee joint appears beautifully complex and functional; the path taken by the sperm tubes from the testes to the penis in mammalian males seems worthy of an engineering award; and the design of the human eye seems a likely candidate for the Nobel Prize in physics. But further analysis shows that these structures are merely the best answers to some unfortunate evolutionary legacies for which we are constantly compensating. In the following pages, we will discuss a small sample of the evolutionary engineering that has occurred in humans as a response to unanticipated changes in our behavior, physiology, and anatomy. An engineer would call many of these designs faulty, and indeed they are, but they are the best that evolution could do, given the material it had to work with at any one time. A good engineer—an intelligent designer—would have done better.

### TWISTS: WEAK IN THE KNEES

In a paper published in 2006 in the journal *Gait and Posture*, C. Owen Lovejoy, an anthropologist at Kent State University, summarizes bad design in human anatomy with the statement, "The mammalian knee suffers from a rather precarious design." To understand the evolution of the mammalian knee, we must journey back to the evolution of terrestrial tetrapods from aquatic fish (see "The Arrival of the Fittest" in chapter 12).

The knee joint had its origin 370 million years ago in the pelvic fin of lobe-finned fishes. Indeed, 360 million-year-old tetrapod fossil bones include the three main leg bones that make up the knee joint in today's four-legged animals: the femur, tibia, and fibula. The first tetrapod animals walked on their toes using a gait called *digitigrade*, in which only the fingers and toes (digits) contact the ground during normal forward movement. Digitigrade motion is still found in most *quadrupeds* (animals that use all four limbs for walking). In mammals, for example, dogs and cats are quadrupeds and their movement is digitigrade. The movement of other quadrupeds is *unguligrade*, in which the knee joint is never fully extended. Such movement is found in mammals like pigs, sheep, and horses. In other groups, such as bears, great apes, and humans, movement is *plantigrade*; species in these groups contact the entire surface of their "hands" and feet with the ground as they move. Digitigrade animals, therefore, have an additional joint (between the "palm" and fingers) that is not found in plantigrade animals and gives them one more hinge for elongation during forward movement. This extra hinge gives digitigrade animals the speed and acceleration we observe in animals like rabbits and cheetahs.

Plantigrade movement evolved primarily to help the foot function as a grasping appendage. In grasping movement in an arboreal (living in trees) setting, the limbs are fixed to a substrate (the tree) most of the time. This arrangement mostly eliminates the need for the foot to elongate as in digitigrade animals. The loss of a hinge during the evolution of plantigrade movement required the knee joint to compensate by increasing its range of motion. This wide range of motion is most pronounced in human *bipeds* (animals that only use two limbs for walking). Today, humans are the only known species that is both plantigrade and strictly bipedal.

During the development of a human embryo, as the knees and elbows begin to develop, they both point backward, recapitulating the elbow and knee orientation in the fossils of adult fish that were making the transition from an aquatic to a terrestrial lifestyle (see chapter 11). However, further along in human embryological development, the knee joint begins to take on its forward-facing orientation, but the elbow stays facing backward. The current state of human knee joint development is a structure so unusual and complex that no other mammalian or primate species can serve as a model for understanding the human knee.

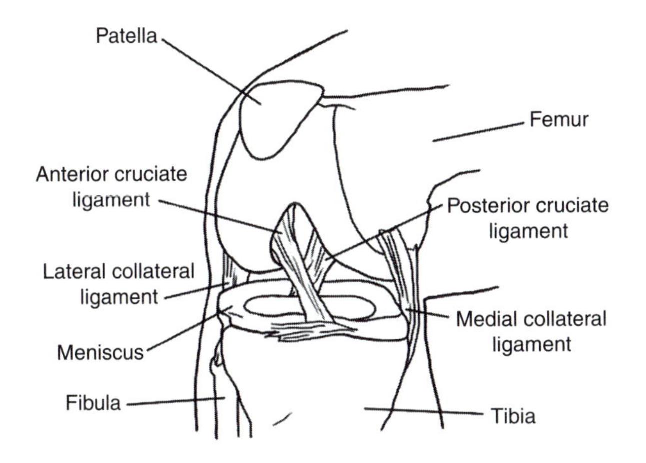


15. The human right knee, partially flexed and viewed from the front and slightly to the inside. Skin and muscle are not included. All four ligaments, the medial and lateral collaterals, and the anterior (ACL) and posterior cruciate ligaments, are subject to damage. The ACL is most prone to damage by natural human activities like sprinting, jumping, and abrupt changes in direction. All ACL tears are complete and require surgery to repair. Drawing by Alyssa Wiener.

The human knee is arguably an engineering marvel. However, it is also the most easily injured joint in the human body. The knee is basically constructed of four bones (see figure 15). The large *femur* (thigh bone) runs from the hip joint down to the knee and rests on the smaller *tibia* (main calf bone). The even smaller *fibula* rides alongside the tibia down to the ankle. The smallest knee bone is the *patella* (knee cap), which sits in front of the knee joint when the leg is extended (as in standing posture). The real complexity is how these four bones are attached to each other and how they interact during knee *flexion* (bending of the leg).

When the human knee moves during flexion, it does not simply bend; rather, the end of the femur in contact with the tibia also rotates to provide the large range of motion required by our bipedal-plantigrade gait. To provide support for this bend-with-rotation motion, the human knee has four ligaments. The *medial collateral ligament* and the *lateral collateral ligament* on either side of the knee joint are attached to both the femur and the tibia and help prevent side-to-side movement of the knee. Two ligaments, the *anterior cruciate ligament* (ACL) and the *posterior cruciate ligament* (PCL), also attach the femur to the tibia, but twist past each other inside the knee joint. (*Anterior* and *posterior* refer to the front and back of the organism.) One ligament, the PCL, runs from the front of the femur to the back of the tibia and twists between the two bones from outside to inside as it goes. This placement keeps the tibia from sliding backward on the femur. The last supporting ligament, the ACL, does just the opposite. It runs from the back of the femur to the front of the tibia and keeps the tibia from sliding forward on the femur. Damage to any one of these ligaments will destabilize the knee joint and cause the knee to buckle one way or the other.

Of all four ligaments, the one most prone to damage is the ACL. The ACL can be torn when the tibia is forced too far forward during jumping, or from a rapid stop or change of direction during sprinting or any form of forward movement. ACL tears are complete; that is, the ligament cannot heal itself back to its pre-injury state and keep the knee joint from buckling. When the ACL is torn, it can be repaired only through surgery. When humans invented sports that involved sprinting and/or changes in direction, injuries to the ACL followed. In fact, today, of all sports injuries that require surgery for repair, injury to the ACL is one of the most common. For example, in a twenty-five-year study of 379 elite French skiers, over one-fourth suffered from a torn ACL while skiing.

The manifestation of the knee joint in humans is a compromise—an evolutionary solution or modification to support the uniquely human plantigrade and bipedal combination. The long lengths of the femur and tibia that give humans the benefit of a long and energy-efficient stride also cause structural weakness: an adaptation with a liability.

### TURNS: THE EVOLUTION OF THE SCROTUM

Many human males (and many other mammalian males, for that matter) who have been hit "below the belt" may have, at that moment of truth, wished they were a harbor seal or an elephant shrew; at least they would have wished it if they had known there is a mammalian alternative to testicular descent. Male shrews in the genus *Elephantulus* leave their gonads (testicles) right where they develop, behind the kidneys and underneath the protective backbone. Harbor seals' testicles move toward their rear flippers and the front of their bodies, but they remain just on the inside of the body wall. As males of many other species mature, however, their testicles descend in front of the pelvic girdle, right through the abdominal wall, and hang dangerously, like pendulums, outside the body in the scrotum. This process is called *testicular descent*, and in some unfortunate individuals, it sometimes stops short of completion. If the testicles fail to descend, the undescended testicle or testicles can cause several problems. Among these problems are abnormal sperm production and infertility, and a 33-fold increase in malignant testicular tumors. If, however, the process is successful, the resulting route that the testicles take forces the *vas deferens* (sperm cord) that carries sperm from the testicles to loop back up and over the front of the pelvis to the seminal vesicle, which produces semen, then to the prostate, which pumps the semen and sperm out through the penis via the urethra. A long and arduous trip, indeed.

Why is testicular descent a characteristic of most mammalian species? How can this process impart any fitness advantage to those species that employ it? The leading hypothesis for external gonads has been that moving the testicles to the outside of the body was an advantageous structural modification as the ancestors of mammals evolved *endothermy* (the production and maintenance of metabolic heat; the same characteristic that for centuries has been erroneously called warm-bloodedness because endothermic animals maintain a body temperature above that of their surroundings). Endothermy eventually created an environment that was hostile to proper sperm development. We owe this problem to our fish ancestors.

The evolution of vertebrate sperm development began in fish. Today's fish have gonads that extend up to their chests and are protected under layers of muscle and skin. In the teleosts, the fish group that includes 96 percent of all living fish species and the group from which humans evolved, *spermatogenesis*, or sperm development, is only an annual process and depends on environmental factors, mostly an annual increase in water temperature. During most of the year, water temperatures are too low to support proper sperm development. After the tetrapod lineage evolved and vertebrates became terrestrial, endothermy soon followed. Endothermy allowed vertebrates to remain active within a much wider range of temperatures and gave them more stamina. Endothermy also reduced the dependence of sperm development on environmental temperature and allowed males to be reproductively active year-round. The body temperatures of mammalian ancestors, however, soon began to exceed the optimum temperature range for efficient and safe sperm development. In humans, this optimum is 2 degrees Celsius below the normal human body temperature of 37 degrees Celsius. At a mere 38 degrees Celsius, mutation rates in sperm DNA reach lethal levels, and the whole process can come to a halt. Something had to change.

In those species whose core body temperature was too hot for sperm development, it would have been more complex to reinvent functional pathways and optimum operating temperatures of the enzymes that drive sperm development than to do a little structural re-engineering. If the tissues in charge of sperm development could be moved just outside the body core, sperm development could continue as it originally evolved. But now, those vital gonadal tissues, the very tissues responsible for organizing and passing on the genetic information of the individual male, hang precariously in harm's way. In fact, many primates (humans, unfortunately, included) direct aggression at a competitor's vulnerable scrotum and its enclosed testicles.

For human males, the descent of the testicles through the abdominal wall and the growth of the scrotum to keep them cool come at a cost. The process creates an opening in the otherwise strong abdominal floor, a weak spot through which the intestines can accidentally protrude as a result of an *inguinal hernia.* This type of hernia, common in human males, is painful and requires surgery to repair. In human females, the ovaries drop only enough so that the fallopian tubes are short and the sperm do not have to travel as far to reach and fertilize a recently released egg. The result is that females have no zone of weakness and rarely suffer from hernias.

Paleozoologists Lars Werdelin and Asa Nilsonne of the Swedish Museum of Natural History argue that the *temperature hypothesis* is not the only possible explanation for the evolution of testicular decent and the scrotum. In the *display hypothesis*, the scrotum, which is brightly colored in some species, signals to females the reproductive readiness of the male. The argument against this hypothesis, however, is that testicular descent is too costly and has too high a noncompletion rate to have evolved merely to create a sexual organ for display. In the *training hypothesis*, the scrotum presents sperm with a hostile environment that, if survived, results in sperm that are trained and ready for the precarious trip that lies ahead. The testicles then end up sending only the best, highest-quality sperm forward. The hostile environment has less to do with the scrotum, however, and more to do with a lack of blood supply, a condition that could have been accomplished without the complex process of testicular descent.

Regardless of which hypothesis rises to the top in the end, if an intelligent designer had truly been involved in protecting sperm development and the testicles, there were other options to consider. For example, if the temperature hypothesis is correct, why not simply bump up a couple of degrees the optimum temperature of the enzymes involved in sperm development, or even lower the mammalian body temperature a bit? Another solution would have been to use the mechanism of counter-current flow already in use in many species, in which heat from blood flowing in one direction (from the body core, for example) is passed to cooler blood in nearby vessels flowing back from the extremities. The cooler blood would maintain the internal and protected testicles at just the right temperature for proper sperm development.

Precisely why testicular descent happened in mammals is unknown. The point is that it happened as a way of balancing cost and benefit. To an individual that gets "hit below the belt," the cost may not be worth it, but to the overall population, it is. The mechanism whereby the temperature problem has been solved, however, is not what a purposeful designer would have done.

## FLIPS: "EYE" TO EYE

As we discussed in chapter 8, Darwin himself could not imagine how the vertebrate eye evolved gradually. Yet, Dan-Erik Nilsson and Suzanne Pelger (see chapter 8) developed a model that tested the hypothesis that the eye could

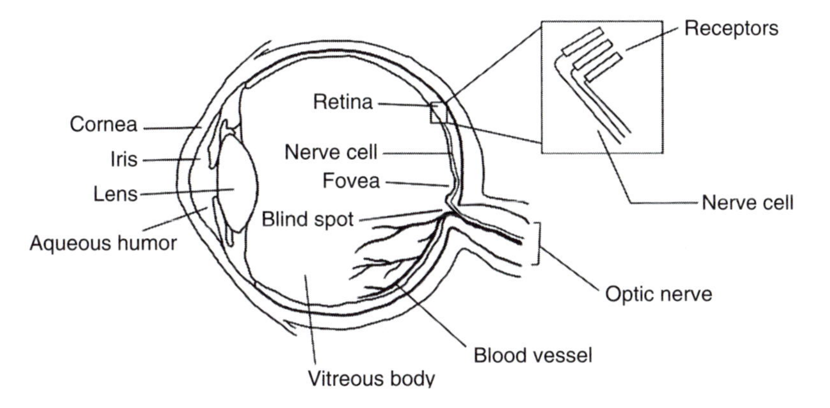


16. Cross section of the human eye, showing that the blood vessels lie on top of the retina and the nerve cells on top of the light receptors. Drawing by Alyssa Wiener.

have evolved in a continuous process from a flat, photosensitive layer of skin to a full camera eye. They found that this process can happen in as few as 500,000 years. We also know from analyzing animal groups and their phylogenetic histories that image-forming eyes have evolved independently in over forty separate animal lineages. And as we discussed in chapter 12, a master control gene for eye development, called *Pax 6*, unites all of these animal lineages to a common ancestor.

We do not deny that the eye is an amazing and marvelous product of evolution, but the eye is far from perfect. Because evolution operates in a rank-order method, the best of available options is what tends to be selected. The eye is the best available option, given its evolutionary history, but we encounter several flaws as we follow a single light ray from outside the eye to the moment the photon energy is absorbed and converted to an electrical impulse and sent to the brain to be interpreted.

A photon first penetrates the *cornea*, a crystal-clear window that performs part of the job of focusing the image (see figure 16). For the cornea to function properly, it is kept moist with tears and well-oiled by glands near the eyelids. The ray next passes through an area behind the cornea filled with a fluid, the aqueous humor, which is responsible for maintaining proper pressure in the eye, and then through the *iris*, the structure that gives eyes their color and controls the intensity of light that enters the inner eye.

Once inside the eye, the ray penetrates the *lens*, which, with the help of several tiny muscles, changes shape to keep the image in focus. The normal vertebrate lens loses flexibility with age, resulting in a condition known as *presbyopia* (Greek for "old man's eye"), or an inability to focus on nearby objects. The eye would be better designed if the lens focused by moving in and out, like a camera lens, and indeed the lenses of cephalopods (squids, octopuses, and their relatives) focus in just that manner. In addition, the crystal structure of the vertebrate lens has two stable states, one transparent and one cloudy. As the normal lens ages, its crystal structure gradually changes to the cloudy state, and the organism becomes blind. The lens in its cloudy state is said to have a *cataract*. Evolution, incidentally, cannot protect a population against cataracts. If cataracts do occur, they likely occur after an individual has already successfully produced offspring, so there is no selection mechanism that can favor lenses that do not form cataracts. Only an intelligent designer could have protected against cataracts.

After the lens, the ray passes through a normally clear, jelly-like substance called the *vitreous body*. With age, the vitreous body becomes more liquid and often detaches from the *retina*, the light-sensitive screen at the back of the eye; pieces of the vitreous body then cast shadows that impair vision or sometimes tear the retina from the back of the eye. Similarly, a forceful blow to the head can easily detach the retina from its underlying support tissue and cause partial or complete blindness.

Finally, the ray strikes the retina, where the image forms. Before reaching the millions of light receptors, called *rods* and *cones*, which make up the retina, the ray first passes through a sea of nerve fibers and a matrix of tiny blood vessels. In vertebrates, this inverted arrangement, in which the rod and cone light receptors sit behind the nerve fibers and blood vessels, instead of being located in front of them, is the best available option considering how the eye develops in a vertebrate embryo.

The two vertebrate eyes develop as outgrowths of the brain, called *optic vesicles*. As the optic vesicles extend away from the brain, their connection to the brain narrows into what will become the optic nerves. The leading tip of each optic vesicle then pulls in on itself (a process called *invagination*), forming the optic cup. From tissue that is trapped inside the cup, blood vessels begin to grow and will become the central artery and vein of the retina. The cells on the outside surface of the cup differentiate into the retina. Therefore, the only option provided by this developmental pathway that involves the formation of two cup-shaped eyes is to have the light-sensitive tissue on the undersurface, or back side—two eyes with inverted retinas.

A second consequence of this developmental pathway and the growth of the eye after birth is that the eye has only a small region of clear vision, the *fovea*, which is not obstructed by nerve fibers and blood vessels. The fovea is necessarily small, in part because it needs to be nourished by the blood vessels, yet to preserve good vision may not be covered by those blood vessels. The fovea is also rich in cones and lacking in rods. Cones provide color discrimination and therefore greater visual acuity, but they are useless at low-light levels. The tradeoff is that the fovea has poor light-absorbing ability because it lacks light-sensitive rods. Such a small region of clear vision with little light sensitivity means that vertebrates with a fovea have poor night vision and are colorblind at low light intensity. That is why you sometimes see a dim light, such as a star, only when you look away from it.

The nerve fibers converge to the main optic nerve, pass through a hole in the back of the eye, and head toward the brain. This concentration of nerve fibers into one central nerve creates a region (called the *blind spot* or the *optic disk*) on the back of the eye, where there are no photoreceptors. The need to keep the optic nerve comparatively small and, therefore, to keep the blind spot comparatively small also necessitates a small fovea: if the fovea were larger, it would hold more receptors and the optic nerve would have to be thicker.

One kind of *glaucoma*, a form of blindness that occurs in the elderly, is apparently caused by damage to the optic nerve in the blind spot, precisely where the optic nerve goes through the hole in the eyeball. The precise cause of the glaucoma is not completely clear, but it may be related to mechanical or other damage that results when the pressure of the fluid inside the eye pushes the optic nerve through the hole and stretches the fibers. Although it is often associated with elevated pressure inside the eye, glaucoma is also found in eyes with normal internal pressure.

Remember Aunt Rita from chapter 8 and her age-related macular degeneration? One variety of macular degeneration, commonly called wet macular degeneration, is caused by bleeding from abnormal blood vessel growth. Because the blood vessels lie on top of the retina, the bleeding obscures vision and often causes nearly complete blindness. Macular degeneration is not rare; in the United States alone, approximately 8 million people over the age of fifty-five suffer from age-related macular degeneration.

Thus, the architecture of our eyes necessitates a very small region of high acuity. Additionally, the passage of the optic nerve through a hole in the eyeball may predispose us to getting glaucoma. Finally, wet macular degeneration is a problem in part because the blood vessels lie on top of the retina. There is, however, no intrinsic barrier to locating the receptors on top of the nerve fibers and blood vessels; it is, in fact, located there in the cephalopod eye. It is located behind the nerve fibers and blood vessels in the vertebrate eye only because of the details of its development from brain tissue.

Contributors to the young-earth creationist website Answers in Genesis (see chapter 3) accept that the retina is inverted. Yet they carelessly argue that the design of the human eye is still "pretty good," any imperfections in the human eye are a product of a biblical event called the Fall, and "the occasional [*sic*] failures in the eye with increasing age reflect the fact that we live in a fallen world." They explain that humans used to exist in an "original physically perfect state" in which "deterioration with age didn't occur." Such arguments hark back to the old-earth creationist Hugh Ross's explanation of supposed 900-year lifespans; there is simply no evidence to support such fantastic claims

### EVOLUTIONARY LEGACIES AND STRUCTURAL RETROFITTING

Many characters in the animal and plant kingdoms (1) probably made sense long ago when habitats and environmental pressures were different, and (2) have been clumsily structured out of existing parts and pathways to perform novel functions. These traits either have persisted because the genes that control their development are still functioning, or have arisen from mutations that created structures that worked well enough to be favored by natural selection. That we find such characters in today's organisms is evidence in favor of descent with modification but against an intelligent designer. The following is a partial list of such characters; the complete list is far too long to include in its entirety here.

*Animals*

1. Genes for teeth development. Birds do not have teeth, but their common ancestor with dinosaurs did. Birds still have the genes for making teeth. Why keep unnecessary DNA around?
2. Goose bumps. Goose bumps (those bumps that appear on your skin when you're chilled) made sense when we had more hair. Body hairs stand more erect when the bumps form, trapping insulating air. After 100,000 years of evolution, most of us don't have enough body hair for goose bumps to make a difference.
3. The panda's "thumb." Pandas are bears and have five fingers that all curl in the same direction for grasping and climbing. They also have what appears to be a sixth finger that works like the opposable thumb of a primate and allows pandas to strip leaves off bamboo stems. The bones for the "thumb" are actually deformed wrist bones, and the muscles that work them have moved from the main part of the hand. It's clumsy, but it works. Still, why not simply design a real thumb for the panda?
4. DNA polymerase III. This enzyme works by building (replicating) new strands of DNA just before the nucleus of a cell divides. It may have evolved before DNA was double-stranded, because it can synthesize only one of the two DNA strands continuously. The other strand (which runs in the opposite direction) must be copied in small fragments.

*Plants*

1. Ribulose bisphosphate carboxylase/oxidase (Rubisco). This enzyme helps all plants attach the carbon atom in carbon dioxide to an organic acid while making sugars in photosynthesis. Rubisco

evolved when there was very little oxygen in the atmosphere. With so much oxygen in the atmosphere today, oxygen competes with carbon dioxide and essentially chokes Rubisco. As a result, when carbon dioxide is unavailable (for example when the pores in leaves are closed to conserve water), plants must consume the very sugars they are trying to generate simply to stay alive. In some plant groups, new enzymes and pathways evolved to mitigate this problem, but they still depend on Rubisco for making sugars.

2. Some flowering plants have separate male flowers and female flowers, yet the male flowers still have female parts without ovules for seed production, and female flowers have male parts that have lost the ability to produce pollen. While the plants can self-pollinate or cross-pollinate with another individual, growing the useless other parts is a waste of energy and material resources.

(see chapter 6). Rather, the design of the eye is "pretty good" because it was jury-rigged by evolutionary processes operating on whatever raw material was available, and the lack of perfection provides strong support against an omniscient, intelligent designer.

## Conclusion

The structures and functions that define a species are products of millions of years of gradual change and modification. But a species cannot simply lose one structure or function in order to replace it with another. In other words, natural selection cannot go down one hill (lose one structure) in order to climb up another (evolve a new structure). The human knee, the mammalian scrotum, and the vertebrate eye are far from perfect, but rather are merely the best evolution can do given the constraints of developmental genes and structures and functions already in place. Ultimately, it's all about balance. We pay a cost, but the overall fitness is better. Perhaps better designs could be envisioned, but evolution has had to work with what it had, finding solutions that improved fitness overall, but with a price.

In the following unit, "The Universe," we temporarily leave biology and take up the question of the age of the earth and the universe. Indeed, in chapter 15, we show conclusive evidence, based on geological and astronomical observations, that the universe is approximately 13.7 billion years old, and the earth, 4.55 billion years old. Further, we explain why certain creationist arguments for a young earth are not consistent with the observations or with their interpretations of those observations. Next, in chapter 16, we discuss modern versions of the argument from design. These arguments use the supposed fine tuning of certain physical constants, such as the charge of the electron and the speed of light, to deduce that the universe was created for us. We conclude that the arguments are circular and that the physical constants may not be so finely tuned after all.

### Thought Questions

1. Think of any human structure or function. If you could re-"design" the structure or function, how would you make it better given the current environmental pressures that humans face?
2. Why do you think so many people are threatened by the fact of evolution and cannot come to terms with it? What features of evolution offend people, whereas other sciences do not?

Part Four

# *The Universe*

CHAPTER 15

# *How We Know the Age of the Earth*

. . . no vestige of a beginning, no prospect of an end.

—James Hutton, Scottish geologist and founder of modern geology

DARWIN WELL KNEW THAT his theory would fall if the earth were not immensely old, far older than the few thousand years that Archbishop Ussher had predicted (see chapter 3). The view that the earth was young was comparatively recent and largely a result of Christians' literalist reading of the Bible. Ancient Greek scholars, such as Xenophanes and Herodotus, recognized that fossils were the remains of organisms that had lived long ago and had somehow been preserved in rocks. Leonardo da Vinci in the late 1400s drew a similar conclusion when he examined the fossils of marine organisms that had been found in the hills of Tuscany in northern Italy; Leonardo deduced, correctly, that the hills had once been covered by the sea and ancient animals had been buried in mud. Later, Nicolaus Steno, an early Danish geologist, demonstrated that the fossils were layered in such a way as to reveal sequential changes and deduced that deeper *strata*, or layers of sedimentary rock, are older than strata above them. Steno enunciated a *principle of superposition* which states that layered rocks had been deposited sequentially, from the bottom up. Later, in 1725, John Strachey, another early geologist, drew maps of rock strata in England.

Following Strachey, other naturalists studied rock strata in England and on the continent, and developed early theories that were consistent with the Noachian flood. Abraham Gottlob Werner, a German geologist, proposed that the flood had deposited the rock sequences and the rocks were then sculpted by the receding waters. His theory has come to be known as *neptunism*, after the Greek god of the sea, Neptune. Neptunism was eventually rejected because it could not account for the presence of igneous rocks; these were presumed to be of volcanic origin and therefore not consistent with a flood theory.

In the late 1700s, James Hutton, a Scottish geologist and naturalist, proposed what is arguably the first modern theory of geology. Unlike Werner,

Hutton did not see the earth as static; that is, he did not see the surface of the earth as having been deposited in its present form. Rather, Hutton argued that the processes we see today—erosion and deposition of sediments, earthquakes, and volcanism, for example—have been taking place since time immemorial and formed the geological features we see today. Hutton's theory is known as *uniformitarianism*, and it slowly replaced Werner's neptunism. Uniformitarianism is clearly in conflict with the hypothesis of a young earth.

Around 1800, William Smith, an English geologist, discovered markedly different fossils in different geological strata and concluded that the earth had been populated by a succession of different floras and faunas. It was unclear at the time whether the marked differences among the species represented gaps in the record or periods in which one set of floras and faunas was relatively suddenly replaced by another. The French naturalist Georges Cuvier showed that the organisms found in fossils are related to present-day organisms but are different species. Adopting a view called *catastrophism*, he argued for a series of catastrophes, such as widespread floods or earthquakes, to explain why new floras and faunas sometimes appeared in the fossil record. Incidentally, nearly all the Eras, Periods, and Epochs of the Phanerozoic Eon, the geologic period between the beginning of the Cambrian Period and the present, were identified and named well before the publication of *On the Origin of Species*. The only changes made after 1859 were subdivisions of the existing periods; no periods were ever rearranged after 1859. Hence, the creationists' claim that the geologic timescale is a fiction based on the "presumption" of evolution is false.

Charles Lyell, the English lawyer and geologist whose influence on Darwin was considerable, took up Hutton's principle of uniformitarianism, though he perhaps interpreted it more literally than Hutton himself had intended. Specifically, Lyell stressed that today's geological processes are continuations of past processes and also asserted that the rates of those processes were precisely the same in the past as they are today. Thus, Lyell presumed not only that rates of erosion and sedimentation were constant, but also that earthquakes and volcanic eruptions have occurred with precisely the same frequency in the past as they occur today. Lyell's multivolume work, *Principles of Geology*, was published between 1830 and 1833. Darwin brought volume 1 with him on the *Beagle*, and it had a profound effect on his thinking. Lyell himself came to accept Darwin's theory of descent with modification only after a struggle, in 1865.

### EARLY ESTIMATES

One of the first methods of estimating the age of the earth was proposed in 1715 by Edmond Halley of Halley's comet fame. Halley reasoned that rivers carry salt (sodium chloride) to the oceans and increase their salinity. If the oceans are well stirred, then measurements of salinity at different times will allow us to calculate the age of the oceans and, by inference, of the earth. Because he had no database of earlier salinity measurements, Halley could not perform his calculation. Later researchers added the hypothesis that the primordial ocean had little or no dissolved salt or other compounds such as sulfates. If the hypothesis was correct, then they could estimate the age of the earth as long as they knew the volume of the oceans and could estimate the mass of the salt that accumulated in the oceans in one year. Specifically, the age of the earth is equal to the mass of salt in the oceans divided by the mass that is added in a single year. John Joly, a professor of geology at the University of Edinburgh, performed this calculation using the best data available to him at the time. Indeed, he refined his calculation to include possible losses of salt. Joly estimated that the earth was not more than 150 million years old. Others performed similar calculations and obtained roughly similar results. Unfortunately, such calculations are almost prohibitively difficult, in part because the influx of salt into the ocean is highly variable over geological times. More important, according to the geologist G. Brent Dalrymple of the University of Oregon, the oceans are now known to be approximately in chemical equilibrium; that is, salt and other chemicals are removed from the water at very nearly the same rate as they are introduced.

Benoît de Maillet, a French naturalist, traveler, and consul to Egypt, was among the first to recognize that the geological record embedded historical information and deduced, wholly incorrectly, that the entire earth had initially been submerged in water. Noting and measuring an apparent decrease in the level of the Mediterranean Sea since ancient times, de Maillet estimated the rate at which water was being evaporated and thereby lost to outer space. He estimated that the tallest mountains began to appear above the water level approximately 2 billion years ago. Unfortunately, de Maillet's observations were geographically too limited, and he did not recognize that the sea rose in some places and fell in others, or that the apparent decrease in sea level in the Mediterranean was largely due to the land's rising, rather than to the sea's falling. Fearing reprisal from the Church, de Maillet arranged for his work to be published posthumously under a pseudonym in 1758.

Other geologists, most notably the American Charles D. Walcott in 1893, calculated the age of the earth by estimating the time required for various geological strata to have been deposited. Walcott divided the process into two stages: the deposition of rocks such as sandstone, and the precipitation of rocks such as limestone from sea water. The calculation was extremely difficult. Walcott estimated the thicknesses of the relevant strata, their net area, the rates of deposition, and the rates of erosion. He then calculated the deposition time for each geological eon and added these deposition times to arrive at a total of 55 million years. Other, similar calculations carried out in the late 1800s and early 1900s yielded values between a few million years and nearly 2 billion years.

## LORD KELVIN'S CHALLENGE

Although early geologists' quantitative methods were flawed, they were correct in assuming that the earth was exceedingly old. Physicists also attempted to measure or estimate the age of the earth. Newton reasoned that a 1-inch sphere of red-hot iron cooled in one hour; he extrapolated, incorrectly, to a sphere the size of the earth and estimated the age of the earth to be 50,000 years. The Comte de Buffon, author of the thirty-five-volume *Histoire Naturelle*, followed Newton's lead and had his foundry cast ten iron balls with diameters ranging from 1/2 inch to 5 inches. He heated the balls until they were white hot and timed them as they cooled through red heat to not visibly radiating to room temperature. From this and similar experiments, Buffon deduced that the earth was approximately 75,000 years old, but considerations of sedimentation rates made him suspect that the earth was in fact much older. Buffon and de Maillet were wrong in their results, but they were wrong in interesting ways, and they pioneered the use of science to answer questions that had previously been the purview of philosophers and theologians.

Lord Kelvin attempted to calculate the age of the earth using a heat-transfer calculation. Kelvin assumed that the earth had been formed as an initially molten ball of rock and is now a uniform, homogeneous solid. He assumed, further, that the surface temperature of the earth was approximately constant at 0 degrees Celsius. The last assumption is not unreasonable, because the surface of the earth is cooled as a result of convection by the atmosphere and radiation into space, and is therefore maintained at approximately the same temperature as the air. The internal temperature of the earth is so much higher than the surface temperature that a small error in the surface temperature is insignificant. Using the heat-transfer theory developed by the French physicist Joseph Fourier, Kelvin calculated the expected temperature of the earth at any given time or distance below the surface.

Kelvin had little data to work with, but he knew from measurements in mine shafts that the temperature increased with depth, as expected. Such measurements, however, were inconsistent, so Kelvin assumed that the temperature increased 1 degree Fahrenheit for every 50 feet of depth. He also needed to know the initial temperature of the earth, which he took to be 7,000 degrees Fahrenheit, an estimate of the temperature of melting rock, which is too high by a factor of more than 3. Finally, thermal conductivity is a measure of the rate at which heat is conducted through a solid, and Kelvin assumed a value equal to the average of the thermal conductivities of three different kinds of rock. He estimated the age of the earth to be approximately 100 million years. Clarence King, the first director of the U.S. Geologic

Survey, refined Kelvin's numbers and his calculation, and inferred an even younger age: 24 million years. Many geologists thought it was much older.

Kelvin's calculation was greatly oversimplified, and he had only very limited and imprecise data regarding the distribution of temperature beneath the earth's surface. Different choices of the initial temperature, the rate of temperature rise with depth, and the thermal conductivity can result in greatly different values for the calculated age. In addition, at least two sources of heat besides the primordial heat—heat released as the liquid core solidified and heat released by radioactivity—were unknown at the time, as was the fact that most of Earth's heat is lost by convection, not conduction.

Kelvin's former assistant, William Perry, recognized that the thermal conductivity of the earth's crust increased with temperature. He further supposed that the earth might have a liquid core and only a thin outer layer of solid rock. He performed a calculation similar to Kelvin's and, guessing that the outer layer might be a few tens of kilometers thick, estimated the age of the earth in the billion-year range. Perry would not have expected us to put great faith in this result; rather, he used it to show that the result of the calculation is critically dependent on our assumptions.

### Radiometric Dating

Kelvin's last pronouncement on the age of the earth was in 1897, the year after Antoine-Henri Becquerel discovered radioactive decay, or radioactivity (see box). Within a very few years, Ernest Rutherford and Frederick Soddy, working at McGill University in Montreal, discovered that the rate at which any radioactive species decays is proportional to the number of atoms in the sample. Thus, if we start with 1 million parent nuclei, then, after a time equal to one half-life, 1/2 million will have decayed; after another half-life, 1/4 million will have decayed; and after a third half-life, 1/8 million. At the same time, after one half-life, 1/2 million daughter nuclei will be generated; after another half-life, altogether 3/4 million daughter nuclei; and after a third half-life, 7/8 million. That is, as the parent nuclei are depleted, the daughter nuclei are augmented. The half-lives of a great many elements have been measured and are known today with an uncertainty of 1 percent or less.

The heat generated by radioactivity would have had little impact on Kelvin's calculation, but Rutherford soon proposed to use radioactivity to measure the age of a mineral directly. Uranium, for example, decays by the emission of an alpha particle, which is a helium nucleus. Suppose that uranium is present as a trace element in a mineral. Any alpha particle that results from the decay of uranium shortly picks up an electron and becomes an atom of helium. Let us suppose that the helium atoms that result from the decay of uranium are trapped in the mineral. For simplicity, we assume further that no helium was trapped when the mineral formed because the mineral was

### RADIOACTIVE DECAY

An atom of a given element consists of a heavy *nucleus* surrounded by lighter *electrons.* The electrons are negatively charged (all electrons have the same charge), and they determine the chemical properties of the element. The nucleus is composed of positively charged protons and uncharged *neutrons.* The protons have the same charge as the electrons, but the sign of that charge is opposite to the charge of the electron. The number of electrons in a neutral atom is equal to the number of protons in the nucleus.

The number of protons in the nucleus is called the *atomic number.* The total number of protons and neutrons in the nucleus is called the *mass number.* Atoms that have the same atomic number are chemically the same and therefore are the same element. Atoms of the same element may have different numbers of neutrons and hence different mass numbers; these are called *isotopes* of that element. Because the number of electrons in the atom is equal to the number of protons in the nucleus, the atomic number indirectly determines the chemical properties of the element, irrespective of mass number. Hence, different isotopes of the same element have almost exactly the same chemical properties. Thus, for example, strontium-86 and strontium-87 are different isotopes of the same element and have mass numbers 86 and 87, whereas rubidium-87 is a different element, even though it has mass number 87.

Many elements have isotopes that are *unstable.* That is, the nuclei of those unstable isotopes randomly emit a particle, such as a helium nucleus (alpha decay) or an electron (beta decay), and *decay* into an isotope of a different element. We cannot predict when a given nucleus will decay, but a large number of nuclei decay on average in a predictable manner. We describe their decay in terms of a number, the *half-life* of the isotope, which is the time required for one half of the nuclei in a given specimen to decay.

The original isotope, the one that decays, is called the *parent,* and the decay product is called the *daughter.* Thus, if rubidium-87 decays to strontium-87, then rubidium-87 is the parent species, and strontium-87 is the daughter species. Atoms of the daughter species are *radiogenic* if they are the result of radioactive decay. Sometimes, we may require an isotope, such as strontium-86, for reference. In this book, we will call that reference isotope the *niece* (or the *nephew* if there is more than one). The reference isotope may or may not be radiogenic.

formed at high temperature, so any helium that might have been initially present escaped.

If our assumptions are valid, then, after one half-life of uranium, the mineral will hold equal numbers of helium and uranium atoms; after two

TABLE 1
*Ratio of Daughter to Parent after Several Half-Lives*

| Number of half-lives | Fraction of parent species remaining | Fraction of daughter species created | Ratio of daughter to parent |
|---|---|---|---|
| 0 | 1 | 0 | 0 |
| 1 | 1/2 | 1/2 | 1 |
| 2 | 1/4 | 3/4 | 3 |
| 3 | 1/8 | 7/8 | 7 |

half-lives, three times as much helium as uranium; and after three half-lives, seven times as much helium as uranium (table 1). Using reasoning such as this, we can derive a simple mathematical formula to calculate the age of the mineral from the measured ratio of helium daughter atoms to uranium parent atoms, and the known half-life of uranium. In 1905, Rutherford estimated the ages of two mineral specimens to be approximately 500 million years. There was evidence, however, that helium had escaped from the material and therefore that the specimens were older than the calculated value. Additionally, isotopes had not been discovered, so Rutherford did not know that uranium has two radioactive isotopes with different half-lives. Thus, his uranium-helium ages were inaccurate for several reasons.

Uranium decays, after a series of intermediate steps, to lead, which is stable and is not volatile like helium. The uranium-helium system was therefore soon replaced by the uranium-lead system, which yielded ages up to 2 billion years. To be sure, there was uncertainty about radiometric dating. Perhaps most important, the initial concentration of the daughter element was unknown; a significant initial concentration would have incorrectly increased the estimated age. Nevertheless, unlike the cooling of the earth or accumulation of sedimentation or salt, radioactive decay provided a stable clock whose rate could be measured in the laboratory. Calculations of the ages of minerals based on radioactive decay are far more reliable than the earlier methods, which have been all but abandoned for quantitative measurements of age.

### Isochron Method

Until now, we have assumed that the initial concentration of the daughter element was zero or at least negligible. If we just use the simple formula for radioactive decay, then a very high initial concentration of the daughter element can render a calculated radiometric date useless. Suppose, for example, that the original specimen contained 1 million parent atoms and 1 million nonradiogenic daughter atoms, that is, atoms that are identical to the daughter atoms but were present in the specimen when it formed. After one half-life of the parent atoms, 1/2 million parent atoms will remain, but there will now be 1.5 million daughter atoms. The ratio of daughter atoms to parent atoms is therefore 3. According to table 1, we expect the age of the specimen to be two half-lives, but in fact its age is only one half-life. Thus, a significant initial concentration of the daughter atoms may result in a significant overestimate of the age of the specimen.

Geophysicists overcome this problem by using an elegant formulation known as the *isochron method* (*iso* comes from the Greek word for "same" and *chron* from the word for "time"; hence, *isochron* means "same time"). Not only does the isochron method get around the problem presented by not knowing the initial concentration of daughter nuclei; it actually permits calculation of that concentration. In addition, an isochron method is self-checking in that it automatically demonstrates whether or not the essential assumptions of the method have been satisfied.

To see how the isochron method works, consider a system that comprises a parent isotope, a daughter isotope, and a nonradiogenic niece isotope. The niece is the same chemical element as the daughter but has different mass number; that is, the daughter and the niece are isotopes of the same element. Specifically, the parent may be rubidium-87, which decays with a half-life of approximately 50 billion years to strontium-87. The niece in this system is strontium-86, which is stable and not radiogenic.

To a geologist, a mineral is a single substance, such as a crystal of quartz or calcite, and a rock is an aggregate, or a blend, of several minerals. The isochron method requires several minerals from the same rock or from rocks that are thought to have formed at the same time. Because each mineral formed under slightly different conditions, they will show different initial concentrations of the daughter element, even though they are incorporated into the same rock. The initial ratios of daughter to niece, however, are the same in every specimen, because the daughter and the niece are chemically identical. As long as their mass numbers exceed 20 or so, the difference between their mass numbers is chemically insignificant, and neither isotope will be preferentially incorporated into a given mineral or diffuse out of it (in chemical terms, the isotopes will not *fractionate*).

Figure 17 is an *isochron diagram*, which we use to illustrate the isochron method. We collect three minerals from the same rock, though in practice we need many more. Consider for a moment one of the three minerals. Let us suppose, for example, that that specimen incorporated 5 million parent nuclei and 1 million daughter nuclei, as well as 1/2 million niece nuclei when it formed. The initial ratio of parent to niece ($P$:$N$) is thus equal to 10 (5 million divided by 1/2 million). Similarly, the initial ratio of $D$:$N$ is 2. Now we begin to construct a graph that shows the ratio of $D$:$N$ on the vertical axis and the ratio of $P$:$N$ on the horizontal axis. In scientific jargon, we say that we are

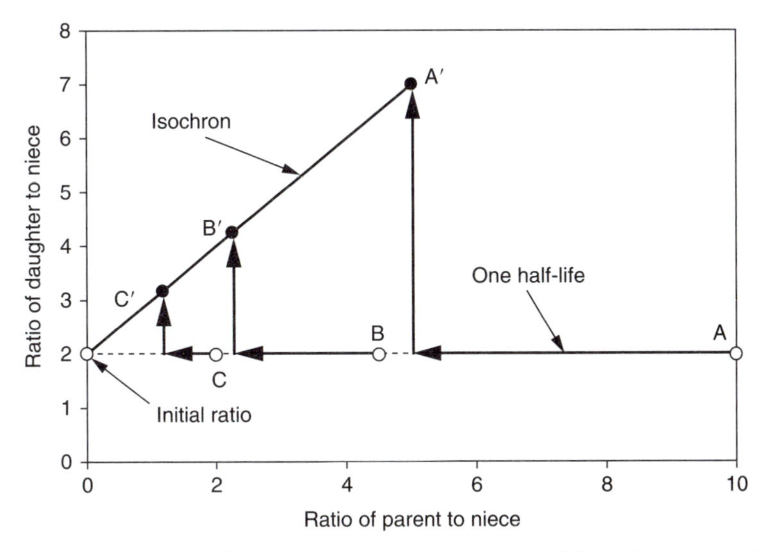


17. Construction of an isochron from three specimens derived from the same rock. The initial ratios are given by the open circles and the measured ratios by the filled circles. In one half-life, the number of parent nuclei decreases by half, while the daughter increases by the same number. The niece remains constant.

*normalizing* $D$ and $P$ to $N$. The specimen we are discussing appears on this graph as open circle $A$.

In one half-life, the number of parent nuclei decreases by half, while the number of daughter nuclei increases by the same number. The specimen that began at $A$ now holds 2.5 million parent nuclei and 3.5 million daughter nuclei (2.5 million plus the original 1 million). $P{:}N$ is thus reduced to 5, as shown by the horizontal arrow, and $D{:}N$ increases to 7, as shown by the vertical arrow. The values 5 and 7, shown by the closed circle at $A'$, are the values we would measure today, if the specimen were exactly one half-life old. Similarly, if the second specimen from the same rock began as the open circle at $B$, today we would measure the values given by the closed circle at $B'$. Finally, the third specimen begins at point $C$ and today is represented by point $C'$. The line that connects the three points is called an *isochron*. We may use the slope of the isochron to calculate the age of the rock from which the two specimens have been taken. Ideally, we must use many more than two specimens in order to ensure that all the points $A'$, $B'$, $C'$, $D'$, and so on lie precisely on a line. The isochron method is thus self-checking in a way that the simple radioactive decay method is not. Additionally, the intersection of the isochron with the vertical axis yields the smallest ratio of daughter to niece atoms, which is the ratio that was present in the specimen when it was formed. Thus, the isochron method yields the initial ratio of daughter atoms to niece atoms, a number that otherwise would have been assumed to be zero.

The uranium-lead system is more complex than rubidium-strontium. Natural uranium has two principal isotopes, uranium-235 and uranium-238. Uranium-238 decays through a series of intermediate steps to lead-206, which is stable. Uranium-235 similarly decays to stable lead-207. The half-lives of the intermediate elements in both series are much shorter than the half-lives of uranium-235 and uranium-238, so they may be ignored, and we may consider that the uranium decays directly to lead with a half-life given by the appropriate isotope of uranium. Uranium-lead dating is especially useful for dating minerals known as *zircons*. These crystals may contain impurities of uranium, but they reject lead, so the initial concentration of lead is zero or nearly zero.

Data are plotted similarly to an isochron, with the ratio of lead-207 to uranium-235 plotted on the horizontal axis and lead-206 to uranium-238 on the vertical axis. The data are plotted alongside a theoretical curve known as a *concordia*. The line that connects the data points is called a *discordia*. The details are complicated, but the intersection of the curve concordia and the line discordia yields the age of the specimen. Alternatively, data may be normalized to the concentration of nonradiogenic lead-204 in the specimen. The ratio of lead-206 to lead-204 is plotted on the horizontal axis, and lead-207 to lead-204 on the vertical axis. As with the rubidium-strontium isochron diagram, the slope of the isochron yields the age of the specimen.

### Archean Rocks and Meteorites

The oldest, or *archean*, rocks may be found in places such as Greenland, where the American continents were separated from Eurasia and Africa due to continental drift and the oldest crust was exposed. Dalrymple cites a dozen or so archean rocks that have been dated by the uranium-lead method, the lead-lead method, the rubidium-strontium method, and others. Many of these rocks are "only" 2.5 billion years old, but others are 3.7 to 3.8 billion years old. The oldest known specimen is a zircon crystal from Canada that is roughly two-tenths of a millimeter across and has been dated as 4.4 billion years. We may be very confident that the earth's crust is at least 4 billion years old. Archean rocks that are less than that age were probably melted and refrozen, so their radiometric clocks were in effect reset.

Meteorites are fragments of stone or iron that fall to earth from outer space. The vast majority of meteorites most probably originated in the asteroid belt, though a few meteorites have come from Mars or the moon. Astronomers think, on theoretical grounds, that the entire solar system, including earth and the asteroid belt, formed at approximately the same time, and there is good evidence based on lead-lead isochrons that their conclusion is correct. Dozens of meteorites have been radiometrically dated, and their age comes out to 4.55 billion years. Similarly, dozens of rocks recovered from the moon have been dated by isochron methods. The oldest are approximately

4.5 billion years old. Rocks that appear younger are still over 3.9 billion years old and were apparently melted or otherwise disturbed by a meteoric impact.

The oldest known archean rock is 4.4 billion years old. The oldest meteorites are 4.55 billion years old, and the oldest moon rocks, 4.5 billion years old. The earth did not form instantaneously but grew to its present size over a period of 50 million years or so by the *accretion* of asteroids, meteoroids, and dust particles captured by the earth in the early solar system. The earth's crust solidified at least 4.4 billion years ago.

### The Age of the Universe

The existence of naturally occurring radioactive elements tells us immediately that the earth is not infinitely old, else they would have completely decayed, and only the daughter elements would remain. To estimate the age of the universe, as opposed to the age of the earth, we turn from geology and geophysics to astronomy and astrophysics. In 1912, the American astronomer Vesto Slipher and others observed the spectra of light from distant galaxies and found that the wavelengths of the light were longer than what we measure in a laboratory. Slipher concluded that the galaxies are moving away from one another and that the spectral shift was due to the *Doppler effect*. Most of us are familiar with the Doppler shift of sound waves: When an ambulance or a police car roars toward us, the frequency of the sound increases; when the car passes us and roars away from us, the frequency of the sound decreases. Because frequency is inversely proportional to wavelength, the wavelength increases as the car moves away from us and decreases as the car moves toward us. The Doppler effect pertains to all wave motion, including light. Thus, when a distant galaxy moves away from us, the wavelengths in its spectrum appear longer than they do in the laboratory. Because red light has a longer wavelength than blue light, we say that the light is *red-shifted*. Some galaxies are moving so fast that their spectra are red-shifted beyond the visible region of the spectrum and into the microwave region.

In 1929, Americans Edwin Hubble and Milton Humason, working at Mount Wilson Observatory in California, measured the distances to a number of galaxies whose redshifts had been measured by Slipher and concluded that more-distant galaxies were receding faster than nearer galaxies. Indeed, they found that the velocity of a given galaxy increased in proportion to its distance from us. The Belgian astronomer Georges Lemaître proposed that the universe had begun a finite time ago as a small volume and has been expanding ever since. If we take the present rate of expansion and include gravitational and other effects, we calculate that the age of the universe is approximately 13–15 billion years.

In 1948, George Gamow and his coworkers, assuming that the universe had once been a relatively small "fireball," suggested that the radiation from that fireball should still be in existence. The astronomer Fred Hoyle, a proponent of the *steady-state theory*, in which the universe is infinitely old, scornfully called Gamow's conception the *big bang*, and the name has stuck. Nevertheless, cosmologists now think that at a certain time in its development, the early universe was a perfect radiator called a *blackbody*. The spectrum of a blackbody is very well known and depends only on its temperature. Gamow estimated that the universe has cooled so much that its temperature should be on the order of 5 kelvins, that is, 5 degrees Celsius above absolute zero. The corresponding blackbody spectrum would then appear in the microwave region of the spectrum. In the early 1960s, Robert H. Dicke of Princeton University and his colleague David Wilkinson set out to detect the radiation, which is now known as the *cosmic microwave background*. In 1965, however, physicists Arno Penzias and Robert Wilson (using a detector invented by Dicke) discovered it more or less accidentally while performing careful radio-astronomy measurements at Bell Laboratory in New Jersey. The cosmic microwave background is considered to be firm evidence in support of the big-bang theory.

In 1989, NASA launched the Cosmic Background Explorer satellite, or COBE. COBE measured the cosmic microwave background and found that its spectrum almost perfectly fit the spectrum of a blackbody with a temperature of 2.7 kelvins. Additionally, COBE showed that the radiation was *isotropic*, that is, the same in all directions. Big-bang theory, however, maintains that the early universe was not perfectly isotropic, else matter would not have clumped and formed stars and galaxies. Accordingly, in 1991, NASA launched the Wilkinson Microwave Anisotropy Probe (WMAP) to look for temperature fluctuations on the order of 1 millikelvin, or one thousandth of 1 kelvin, and with better spatial resolution than COBE. The results supported the big-bang theory with remarkable accuracy and also are consistent with the observed distribution of matter in the universe today. The results demonstrate that the universe is 13.7 billion years old, with an uncertainty of approximately 0.2 billion years.

Finally, the age of the universe according to COBE and WMAP is corroborated by astronomy and astrophysics. We will not go into detail but will just note that astrophysicists have what appear to be good, quantitative theories regarding the birth and death of stars. These theories are based on a combination of observation and calculation. The oldest stars in the Milky Way are thought to be between 12 and 14 billion years old, but the uncertainty is considerable, perhaps 2 billion years. In addition, astronomers can perform spectroscopic measurements and estimate the relative concentrations of different radioactive elements in a star. Using, for example, the ratio of thorium-232 to uranium-238, they can perform a calculation similar to radiometric dating and estimate the age of the star. The calculation is complicated by the fact that nuclear synthesis continuously creates the elements inside the star, and astronomers have to make certain assumptions about the rate of nuclear synthesis. Astronomers have discovered a number of very old stars, which have relatively high concentrations of thorium and uranium, as measured from their spectra. The oldest such star known has been dated to 13 billion years, a date consistent with the 13.7 billion years given by WMAP for the age of the universe.

### Creationist Quibbles

The present estimated age of the earth is based on one hundred years of research into radioactive decay and radioactive dating. Old-earth creationists and intelligent-design creationists obviously have no quarrel with a 4.5-billion-year-old earth nor with a 13.7-billion-year-old universe. How do young-earth creationists respond? Generally, in one of two ways: First, they resurrect long-rejected arguments such as those based on sedimentation rates and on the accumulation of salt in the ocean. Second, they argue that the scientific methods are flawed. Thus, the fundamental constants may not be truly constant, so that the calculations based on the cosmic microwave background, for example, are not relevant; or radioactive dating methods are based on faulty assumptions.

#### *Accumulation Methods*

Chris Stassen, in his Talk Origins article, lists four main methods used by young-earth creationists to date the earth or the earth-moon system. Three of these are accumulation methods: the accumulation of dissolved salts and other chemicals in the oceans, helium in the atmosphere, and meteoric dust on the moon. The accumulation methods are based on influxes but ignore outfluxes. As we noted above in connection with Edmond Halley, the oceans are very nearly in chemical equilibrium. That is, dissolved salts and other compounds are removed from the oceans at approximately the same rate as they flow in, so calculations based solely on influx are fatally flawed. Helium is generated by radioactive decay. It escapes from the upper atmosphere, however, and the concentration of helium in the atmosphere is roughly constant. Creationists have also revived the erosion and sedimentation methods that we discussed above, but they have had no greater success than their forebears, who had to work without our modern understanding of geology and geophysics.

Young-earth creationists further argue that if the earth and the moon are 4.5 billion years old, then meteoric dust upon the surface of these bodies would have accumulated to a depth of about 5 meters. That dust would have been subjected to erosion on the earth but not on the moon, so, creationists say, the moon would have been covered by a thick layer of dust if it were as old as it is thought to be. Their estimate of the rate of influx of meteoric dust, however, is based on a measurement made in 1957 in the earth's atmosphere and, as was later found, contaminated by dust of terrestrial origin. The actual rate of influx of meteoric material is now known from satellite measurements and is 500 times less than creationists claim. Because of the lower gravity of the moon, it is approximately a factor of 2 less on the moon than on the earth, so the creationist estimate of depth is in error by a factor of approximately 1,000. To their credit, some young-earth creationists have yielded on this point, yet we still sometimes see the argument being made.

*Decay of Earth's Magnetic Field*

Creationists also use the supposed decay of the earth's magnetic field to estimate the age of the earth. The earth's magnetic field is somewhat crudely approximated as a *dipole* field, that is, the field that we would detect if the earth were essentially a bar magnet, with a north and a south pole, suspended in space. One way to measure the strength of the magnetic field is in terms of its *dipole moment*, and creationists claim that the dipole moment has been decreasing steadily since it was first measured in the 1800s. The calculation arbitrarily assumes that the decay is exponential (that is, can be described by a half-life, like radioactive decay) and has been steady since the creation of the earth. Equally arbitrarily, the calculation assumes that the magnetic field at the time of creation was equal to the magnetic field observed in a highly magnetic star. The age of the earth is thereby deduced to be about 10,000 years.

Modeling the earth as a simple bar magnet is a gross oversimplification. But the major flaw in the argument is that the earth's magnetic field is known to have oscillated over geological times. Magnetic field reversals were discovered in the mid-1960s by measuring the direction and intensity of the magnetization of lava flows up to approximately 4 million years old. The lava had been magnetized when it was liquid, and the magnetization was frozen into the lava when it solidified. Shortly afterward, scientists found evidence that material had welled up from the floor of the Atlantic Ocean (see chapter 4). A magnetic map of the ocean floor turned out to be "striped," with the stripes running generally north to south. The frozen magnetic fields were oriented in opposite directions in adjacent stripes. Finally, the pattern of the magnetic fields was symmetrical about a centerline that ran from north to south and roughly bisected the Atlantic Ocean. As the continents drifted apart, molten rock constantly welled up from the earth's interior. The earth's magnetic field magnetized the molten rock, and the direction of magnetization was frozen into the rock after it solidified. The magnetic stripes show conclusively that the earth's magnetic field has changed direction over geological times. The earth's magnetic field has not decayed steadily since the creation, but has oscillated; any calculation that assumes an exponential decay of the magnetic field is flatly wrong.

*Change of Fundamental Constants*

Young-earth creationists also argue that the fundamental constants may have changed with time. Physicists have thought of that possibility, and there is good evidence that they have not changed or have changed very little. Astrophysicists, for example, can measure the spectra of chemical elements billions of years in the past. Because of the quantum-mechanical nature of the atom, isolated atoms emit or absorb light only at discrete wavelengths. These wavelengths are called *spectral lines*. Many spectral lines, called doublets, are split into two lines that are close together in wavelength. The magnitude of this split is related to a dimensionless quantity known as the *fine-structure constant*. The fine-structure constant is important in other areas of physics, but it was discovered in connection with spectral doublets.

The spectra of distant astronomical objects are red-shifted (see earlier in this chapter) and have to be observed with radio telescopes. If the value of the redshift is known, then the corresponding spectra measured in the laboratory may be calculated and compared with the spectra observed in astronomical objects. In particular, astrophysicists have measured the absorption of light by atoms and even molecules in distant gas clouds, and compared the splitting of several doublets with laboratory values. In 10 billion years, the fine-structure constant has changed by no more than one part in 100,000, not nearly enough to change our calculation of the age of the universe by a significant factor. More recently, Michael Murphy of the Swinburne University of Technology and his colleagues have performed similar measurements on the spectra of ammonia and other organic molecules in gas clouds and have concluded that the ratio of the mass of the proton to the mass of the electron has changed less than a few parts per million over more than 5 billion years.

Uranium-235 is the uranium isotope involved in fission reactions; natural uranium contains approximately 0.7 percent uranium-235. Uranium-235 decays approximately 6 times faster than uranium-238; that is, its half-life is approximately 6 times shorter. Thus, 2 billion years ago, the ratio of uranium-235 to uranium-238 was approximately 3 percent, which is high enough to fuel a fission reactor. Indeed, in the 1970s, French scientists discovered evidence of an accumulation of uranium dense enough to have fueled a natural reactor that ran for several hundred thousand years, approximately 1.7 billion years ago. The concentrations of isotopes of samarium and europium have been measured, and the physicist Alexander Shlyakhter used these measurements to show that the fine-structure constant and the strength of the force that holds the nucleus together are the same today as they were 1.7 billion years ago.

The fine-structure constant is a ratio involving three fundamental constants: the charge on the electron, the speed of light, and a number, Planck's constant, that relates to the quantization of energy in quantum mechanics.

The astrophysical evidence argues that those three constants, the masses of the proton and the electron, and the force that holds the nucleus together have been constant or very nearly constant for billions of years.

*Polonium Halos*

Halos, in this context, are visible defects formed in rocks by the radioactive decay of specks, or *inclusions*, of radioactive impurities such as uranium and polonium. Polonium atoms, for example, emit alpha particles when they decay. When they are emitted, the alpha particles have a very high velocity and do not interact much with the surrounding rock, because they are moving too fast and, in a sense, pass by the atoms in the rock before those atoms have a chance to respond. Gradually, however, the alpha particles slow down and cause visible damage over a short distance just before they stop. Since a great many alpha particles are emitted, and they are emitted in all directions, the result is a spherical shell of damaged rock surrounding the polonium inclusion. If the rock is polished or sliced into a thin section and examined under a low-power microscope, the spherical shell may be seen as a circle centered about the inclusion. That circle is known as a polonium halo. Polonium has eight isotopes, and each emits an alpha particle with a different velocity from the others, so sometimes several halos with different diameters surround a single inclusion. Of the eight isotopes of polonium, only three cause polonium halos in igneous rocks.

Polonium halos may be found in igneous rocks that formed over millions of years. Robert Gentry, a physicist with the Oak Ridge National Laboratory and a proponent of a young earth, has argued, to the contrary, that the presence of polonium halos in igneous rocks is evidence that those rocks were created in a short time. Specifically, Gentry notes that polonium is one of the decay products of uranium, but polonium halos are found in rocks that do not contain uranium. Further, the half-life of polonium is short; none of the three isotopes associated with polonium halos has a half-life longer than 140 days. If the rocks had solidified over millions of years from molten rock, or *magma*, says Gentry, the polonium would long ago have decayed, and there would be no polonium inclusions in the solid rock. Halos would not form in liquid magma, so the rock must have been created in a solid form and in a short time.

Other scientists disagree with this view. An amateur geologist, John Brawley, writing on the Talk Origins Website, argues that the halos may not be polonium halos but rather are radon halos. He provides evidence of a halo that forms along the length of a crack and argues that that halo resulted when radon gas filled the crack and decayed. Lorence G. Collins, now a professor emeritus of geophysics at California State University at Northridge, also takes issue with Gentry's analysis. First, Collins notes that the rocks that contain polonium halos were formed by the replacement of one mineral by another; rocks that were formed directly from cooling magma do not exhibit polonium halos. If all the rocks had been created at the same time, we would expect to find polonium halos in both kinds of rock, but we do not. In addition, polonium halos are found exclusively in minerals found in uranium-rich areas. These minerals contain microscopic fractures through which water can flow, and in mica, for example, polonium halos are aligned along such fractures. Further, the three polonium isotopes responsible for the polonium halos are all decay products of uranium-238; oddly, polonium isotopes that are decay products of uranium-235 and uranium-232 are not implicated in the generation of polonium halos, but they would be if the rocks had been created in a short-enough time. Finally, Collins notes that radon-222 is one of the decay products of uranium-238. Radon-222 is a precursor to all three of the polonium isotopes associated with polonium halos. Its half-life is several days, which is thousands to millions of times longer than the half-life of any other isotope of radon.

Collins argues that polonium halos formed when water that contained dissolved radon gas flowed through fractures in the rocks. By the time the water entered the cracks, only the radon-222 remained, because of its long half-life. When the radon-222 decayed, polonium compounds precipitated onto favorable sites in the rocks. There is no need to invoke supernaturalism or abrupt appearance to explain polonium halos.

*Excess Helium*

D. Russell Humphreys and his young-earth creationist colleagues at the Institute for Creation Research and the Los Alamos National Laboratory claim to have found zircons that have more trapped helium than is expected. They measured the concentrations of helium, uranium, and lead in a number of specimens that Robert Gentry provided them. They used a number of questionable assumptions to estimate the fraction of radiogenic helium remaining in the zircons. They developed a mathematical model based on standard diffusion theory and claimed that the zircons contain too much helium for them to be 1.5 billion years old and that the data support a young earth. This study is one of several performed under the creationist RATE Project, for Radioisotopes and the Age of the Earth.

Humphreys admits that the calculated uranium-lead ages of his specimens are about 1.5 billion years but claims that the specimens contain too much helium to be more than 6000 ± 2000 years old (an estimate that conveniently encompasses Bishop Ussher's value [see chapter 3]). Humphreys can account for the excessive helium, he says, with his own model and criticizes geologists for their supposedly unjustified attachment to uniformitarianism. Humphreys's model, however, invokes an unusually high radioactive decay rate to supposedly explain the high helium concentrations. Indeed, Humphreys appeals specifically to the Bible to justify his assumption; in short, Humphreys invokes a miracle to establish what scientists call the *initial conditions*, or the state of his minerals at the beginning of his calculation. The assumption of a very high decay rate conveniently invalidates the uranium-lead age of the specimens. Humphreys leaves unexplained why the heat from the sudden decay did not melt or even vaporize the zircons and their surroundings. Indeed, the presence of helium in zircons actually refutes the claim of accelerated radioactive decay, because helium would not accumulate in very hot zircons and rocks, even if they did not melt.

Kevin R. Henke of the University of Kentucky has identified a great many flaws in the work of Humphreys. Many of these are very technical and do not bear repeating here, but Henke identifies a number of instances where Humphreys may have misidentified his specimens and fudged his data. The most important flaws in Humphreys's work, however, may be conceptual. He made the uniformitarian assumption of constant temperature over time, and he did not investigate the possibility of helium contamination.

The diffusion of helium out of a zircon is described by a constant known as the *diffusion coefficient*. The higher the diffusion coefficient, the greater the rate of diffusion out of the mineral will be. The value of the diffusion coefficient is not well known, so Humphreys had it measured. Unfortunately, the measurement was performed in a vacuum; the rate of diffusion in zircons at the high pressures found beneath the surface of the earth can easily be 1 million or even 1 billion times less than in a vacuum. Additionally, the diffusion coefficient depends very strongly upon temperature; a relatively small change in temperature may well change the diffusion coefficient by a factor of 10. Henke argues that the temperature of the rock in which the zircons were found probably varied by at least 100 degrees Celsius in the recent past.

Helium that is not derived from within the specimen is known as *extraneous helium*. Humphreys has assumed that all of the helium in his zircon specimens was generated by radioactive decay within the specimens. The earth's crust normally contains a certain concentration of helium, however, and helium can diffuse both ways, into or out of a mineral. It is thus plausible that at least some of the helium in Humphreys's specimens is extraneous. Extraneous helium has been found within a few kilometers of the Fenton Hill, New Mexico, field site where Humphreys collected his specimens. Further, uranium deposits are found within fractures in the rock cores at Fenton Hill; these indicate that fluids rich in uranium and its extraneous helium once flowed through these rocks.

The isotope helium-3 is not a radiogenic isotope; that is, it is not and cannot be the result of a radioactive decay within the specimen. Henke argues that Humphreys therefore should have looked for (extraneous) helium-3 in his specimens; had he found any, he could be fairly confident that at least some of the helium-4 in his specimens was also extraneous. Humphreys argues that the helium in the zircons cannot be extraneous because the surrounding biotites contain too little helium to contaminate the zircons, but he did not test his contention by looking for helium-3. Further, he improperly assumes that the helium concentrations over the entire history of the biotites have been the same as the current value. If the rocks had been periodically contaminated by extraneous helium in the past, Humphreys's uniformitarian assumption would not be true. Extraneous helium can easily pass into and out of the permeable biotites at an elevated temperature but become trapped in the far less permeable zircons when the temperature falls. Humphreys has evidently taken none of the precautions that careful scientists take to ensure that their measurements and conclusions are accurate.

## Conclusion

The antiquity of the earth was apparent to many observers, from Xenophanes to Leonardo, but an overreliance on the biblical account caused others to think that the earth was only a few thousand years old. The earliest geologists, at the dawn of modern science, constructed theories that were consistent with the biblical account. As evidence accumulated, geologists conjectured that the earth was many millions if not billions of years old. Calculations based on sedimentation and accumulation rates, unfortunately, are prohibitively difficult because of the variations of those rates with both time and location. Calculations such as Lord Kelvin's are likewise inconclusive because even today there are too many unknowns to allow an accurate calculation. By 1900, geologists thought that the earth was very much older than had previously been thought.

The discovery of radioactivity and the development of radioactive dating provided an accurate clock with which the ages of rocks and minerals could be dated. Nearly one hundred years of activity measuring radioactive decay rates and developing radiometric dating techniques have shown without doubt that the earth is approximately 4.55 billion years old. The universe likewise was initially dated by observation of the cosmic red shift. Subsequently, a very precise comparison between the predictions of big-bang theory and careful measurements of the cosmic microwave background allowed us to calculate the age of the universe as 13.7 billion years. These values are not in doubt, though young-earth creationists struggle to find auxiliary hypotheses (see chapter 4) with which to impeach them.

In the next chapter, we will examine the contention that the universe has been designed for life. We will conclude that there is no credible evidence to support that contention and that much of the argument is circular. It is more likely that life evolved to live in the universe than that the universe was intended for life.

### Thought Questions

1. In Lord Kelvin's time, the physicists probably wondered what the geologists were doing wrong. Can you think of any other examples where someone's training got in the way?
2. Distinguish between uniformitarianism and catastrophism, and explain whether either is wholly correct and why. Can you think of any catastrophes in the sense that Cuvier might have used the term?

CHAPTER 16

# *Is the Universe Fine-Tuned for Life?*

> . . . the Anthropic Principle does not assert that our existence somehow *compels* the laws of physics to have the form they do, nor need one conclude that the laws have been deliberately designed with people in mind. On the other hand, the fact that even slight changes to the way things are might render the universe unobservable is surely a fact of deep significance.
>
> —Paul Davies, physicist and writer,
> Arizona State University

THE FUNDAMENTAL physical constants are the charge of the electron and the proton, the mass of the electron, the mass of the proton, the gravitational constant, and others. Every electron and every proton has the same mass and charge as every other. Likewise, the charge of the electron is exactly equal to that of the proton but opposite in sign; that is, the proton carries a positive charge, and the electron a negative charge. On the other hand, the mass of the earth and the acceleration due to gravity are not fundamental constants because we can, at least in principle, express them in terms of one or more of the fundamental constants. The values of the fundamental constants themselves cannot be expressed in terms of any other constants, as far as we now know; that is why we consider them fundamental.

No one knows why each electron or each proton is identical to all others. The Oxford theologian Richard Swinburne, however, uses the fact to develop an interesting variation on William Paley's argument from design (see chapter 7): If he found a large number of identical coins in an archaeological dig, Swinburne says, he would naturally infer a common origin and a common creator. Thus, he infers that identical electrons and protons must also have been designed. The argument is no better than Paley's: The coins are obviously artificial, whereas the electrons are not obviously artificial. Swinburne may have correctly inferred a common origin of electrons and protons, but he has no basis to infer a creator. His reasoning is, like Michael Behe's, another

God-of-the-gaps argument (see chapter 8); it will fail if we figure out why all electrons and protons have the same charge.

Some scientists argue that each fundamental constant has to have precisely the value we observe, or else life would be impossible: change a single constant by a few percent, they say, and we will not be here. For example, the chemical elements are formed in the interiors of stars, and most of the heavier elements could not have formed in abundance unless carbon had had an unusually high probability of being formed. Without carbon, there would presumably be no life, and, indeed, the astronomer Fred Hoyle discovered why carbon was produced in abundance by reasoning backward from the fact of our existence. Supporters of fine-tuning go farther, however, and argue that carbon has such a high probability specifically to ensure that life will develop.

Such reasoning is sometimes called the *anthropic principle*, which has several variants. The strong form of the anthropic principle says that the fundamental constants have the values they do precisely so that intelligent life will evolve. The strong form implies intentionality or purpose. We consider it a circular argument: The fundamental constants have the values we measure in order that intelligent life will evolve, and we know it because we are here to measure them. The weak form of the anthropic principle is innocuous and states merely that we are here because the universe has the right properties; we may deduce from the weak form that we are here now because that is how long it takes for stars with planets to form and for multicellular life to evolve. Thus, we should not be surprised that the universe is 10 or 20 billion years old now. But that is a long way from saying that the properties of the universe were specifically designed for life (and human life, at that).

Supporters of the strong form confuse *prior probability* with *posterior probability*, that is, confuse the probability of an event before it happens with the probability after it happens. Assume, for example, that Babe Ruth has hit a home run that broke the windshield of Grandfather's car. The odds that the ball would have hit any given car were fairly small. There were a great many cars in the parking lot that day, so it was very probable that the ball would have hit some other car, if not Grandfather's. Finally, it is unlikely that the ball, if it hit Grandfather's car at all, would have hit the windshield. But it did hit the windshield of Grandfather's car, and the odds that it did so, after the event, were 100 percent. We have to be very careful when we apply statistics to a sample that contains only one element (see also chapter 8).

Similarly, we have no reason to assume that the universe got its fundamental constants by drawing their values out of a hat, so we have no obvious justification for applying probability theory to the values of those constants. The universe is here and has the constants it has, with odds equal to 100 percent. It may have got those constants during the big bang by a process that we do not yet understand. More specifically, there is no reason to assume that the fundamental constants are independent of each other; if they are not, then simple probability theory does not apply. Maybe the universe assembled itself in the way it did because of a series of causal relationships, just like the eye or the Boeing 747.

### Toy Universes

The astrophysicist Victor Stenger notes that supporters of the anthropic principle fixate on one fundamental constant at a time and claim that its value is fixed by the requirement that the universe support life. Stenger argues that it is not fair to examine the implications of altering only one constant at a time; this is especially so if the constants are somehow related. Rather, says Stenger, to speculate on how things would have turned out with changes in the constants, we must look at all the fundamental constants, vary them randomly, and see what universe develops. If a substantial fraction of the universes have the properties necessary to support life, then our universe is not special, and the anthropic argument is undermined.

Heavy elements are formed in the interiors of stars. Iron and heavier elements are formed in supernovas. Let us assume that stars are necessary for life and that only long-lived stars can form the heaviest elements. There are a lot of fundamental constants, and Stenger has chosen four for examination: the masses of the electron and the proton, and the strengths of the electromagnetic and strong nuclear interactions. First, he used well-known formulas to calculate the lifetime of a star in terms of the four constants. Then he varied the values of the constants randomly over 10 orders of magnitude (an order of magnitude means a factor of 10, so he has varied each constant from 100,000 times its measured value in our universe to 1/100,000 times its measured value). Using the randomly chosen values and the formulas he derived, Stenger calculated the lifetimes of stars in one hundred "toy universes." The number of universes with a given stellar lifetime is plotted on the vertical axis in figure 18. The horizontal axis represents the calculated lifetime of a star and is plotted on a logarithmic scale.

Approximately half of Stenger's toy universes support stars with lifetimes in excess of 10 billion years, and a majority in excess of 1 billion years. Our sun is estimated to be about 4.5 billion years old, so we may assume that 5–10 billion years is long enough for intelligent life to develop. Because more than half of Stenger's universes displayed stars with lifetimes in excess of 5 billion years, his calculation suggests that there is nothing special about our universe and its fundamental constants. The fine-tuning argument may be in error because its proponents do not examine enough alternatives.

### Goldilocks and the Third Planet

Copernicus taught us that the earth was not special, not the center of the universe. Some creationists are trying, in a sense, to return it to that position.

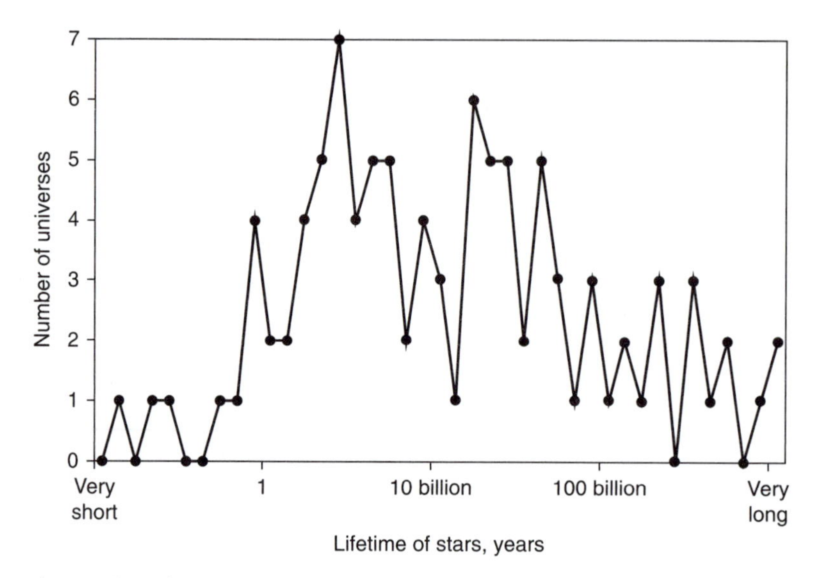


18. The number of toy universes that displayed stars with a given lifetime. The horizontal axis is plotted on a logarithmic scale. Over half the stars have lifetimes greater than 5 billion years and therefore are arguably long-lived enough to support life. After Stenger, Victor. 1995. *The Unconscious Quantum: Metaphysics in Modern Physics and Cosmology*. Amherst, New York: Prometheus.

They argue that the earth is uniquely located in the cosmos to support life, indeed, to support intelligent life. Their argument is flawed from the outset inasmuch as it assumes that "life" means "life based on carbon and water." But let us persevere.

Briefly, the earth is the right distance from the sun: If it were nearer, it would be too hot, and the water would boil off, whereas, if it were farther, surface water would freeze. If the sun were substantially smaller, the earth would have to be closer in order to achieve the right temperature. It would then, like Mercury, always face one hemisphere toward the sun and therefore would be too hot on one side and too cold on the other (the temperate zone in between would probably be very dry because any water would have condensed on the dark side). Additionally, the earth is protected by the giant planets, Jupiter and Saturn, whose gravity sweeps up passing comets and ejects them from the solar system, so they will not crash into the earth. Likewise, the earth is not too near the center of the galaxy, where harmful radiation would militate against the existence of life (again, life as we know it). The earth, in short, is in the Goldilocks zone—neither too hot nor too cold, neither too near nor too far, but rather just right.

The astronomer Guillermo Gonzalez, while on the faculty of Iowa State University, argued that the earthly environment is not only habitable (to us)

Santa Clara County Library
District
408-293-2326

Checked Out Items 4/22/2017 17:53
XXXXXXXXXX8169

| Item Title | Due Date |
|---|---|
| 1. Why evolution works (and creationism fails) 33305218637126 | 5/13/2017 |
| 2. You bian Zhongguo 33305216574933 | 5/13/2017 |
| 3. The language of God : a scientist presents evidence for belief 33305211676493 | 5/13/2017 |
| 4. Creation : how science is reinventing life itself 33305228346346 | 5/13/2017 |

No of Items: 4

Amount Outstanding: $16.00

24/7 Telecirc: 800-471-0991
www.sccl.org
Thank you for visiting our library.

**Santa Clara County Library District**

408-293-2326

**Checked Out Items 4/22/2017 17:53**

**XXXXXXXXXX8169**

| Item Title | Due Date |
| --- | --- |
| 1. Why evolution works (and creationism fails) 33305218637126 | 5/13/2017 |
| 2. You bian Zhongguo 33305216574933 | 5/13/2017 |
| 3. The language of God : a scientist presents evidence for belief 33305211676493 | 5/13/2017 |
| 4. Creation : how science is reinventing life itself 33305228346346 | 5/13/2017 |

No of Items: 4

**Amount Outstanding: $16.00**

24/7 Telecirc: 800-471-0991

www.sccl.org

Thank you for visiting our library.

but also *measurable*. By measurable, he means that we can successfully make measurements of physical quantities such as the laws of motion, our position in the universe, the age of the universe, and so on with relative ease. Again, we are dealing with a sample with only one element—we do not know whether physical laws are hard or easy to discover, because we have no other planets or universes with which to compare. We do know, however, that the deeper we go, the harder it becomes to infer a law: Quantum mechanics and relativity were harder to deduce than Newton's laws, which in turn were harder than Galileo's. Even so, it took 2000 years for Galileo to overturn Aristotle's "laws" (and slightly longer to progress from earth, air, fire, and water to chemical elements). Maybe things are easily measurable only in hindsight and are not at all measurable at the deepest levels.

The claim that a Goldilocks earth that implies a creator may be influenced by experimenter bias and selecting only those attributes that support their hypotheses. Even if they are not, their argument is a specific example of the strong anthropic principle, which we have already seen to be circular. It is used to infer that the earth is somehow special and therefore has been designed with intelligent life in mind. But let us look more closely at the argument.

There are approximately $10^{23}$ stars in the observable universe (the whole universe may be much bigger, possibly even infinitely big; we do not know). There are more stars in the observable universe than there are grains of sand in the Sahara, Gobi, Mojave, and all other deserts combined. Thus, using essentially the explanatory filter (see chapter 8), supporters of a special earth claim that the existence of the earth as an oasis of life is so improbable that it must be purposeful. Hence, they infer a designer.

Let us examine the special-earth theory from the perspective of reaction formation. In chapter 5, we noted an astrologer's claim, in effect, that if a person had five planets in Aries, then either he was meek or he was not. Such a claim, we further noted, is not testable, since *everyone* is either meek or not. What if the supporters of the special-earth hypothesis had found a universe that was teeming with life? That is, what if they thought that conditions favorable for life were common? Why, then, they would have concluded that the universe was designed for life.

In short, if life is uncommon, they conclude that the earth (at least) was designed for life. If life were common, they would conclude that the earth (as well as the rest of the universe) was designed for life. Thus, if life is either common or uncommon, the earth must have been designed to support life. The logic is no better than the astrologer's, because the hypothesis that the earth was designed for life is not testable. Like astrology, the special-earth theory is pseudoscience.

In the 1930s, Lemaître and others proposed that the universe was expanding (chapter 15). Some reputable theists, as well as creationists, nowadays use the expanding universe and the big bang theory to suggest that the universe had a beginning, and they infer from this a purposeful creator. The expansion of the universe, however, implies that the universe has a finite age and will someday die out as the nuclear fuel in the stars is used up (what is sometimes called the heat death of the universe). The historian of science Helge Kragh of the University of Aarhus in Denmark notes that in the 1930s, when big bang theory was not so well established, the physicist Robert Millikan and others argued on philosophical and theological grounds in favor of a steady-state universe. Partly, they did not like the heat death of the universe. Millikan had earlier been the first to measure the charge of the electron, using his famous oil-drop experiment. In the face of evidence from the cosmic red shift, he could not deny that the universe as a whole was expanding, but instead was among those who postulated the continuous creation of matter to maintain a steady state; his conception of a steady-state universe was later taken up by the astronomer Fred Hoyle. When the evidence in favor of the big bang theory and against continuous creation became overwhelming, theological arguments favoring the steady-state universe vanished. Theology and philosophy, however, are extremely flexible and can be made consonant with either theory, just as the special-earth hypothesis can be made consonant with a universe that is either teeming with life or in which life is exceedingly uncommon.

### Alternate Universes

It is theoretically possible, even plausible, that there are a great many universes like our own. For example, maybe the universe—or *multiverse*—is vastly larger than the observable universe, possibly even infinitely large. In this view, the observable universe is but one of many that lie side by side but are inaccessible to one another because of limitations imposed by the speed of light. Such a multiverse is analogous to a weather map that shows regions of high and low pressure that are relatively isolated from one another in the sense that a flying insect that remains near the center of one of these regions is wholly unaware of any other regions. In this view, we happen to live in one of those universes that happens to support life; possibly other universes have properties that do not allow them to support life. There is thus nothing special about our universe, except that it happens to be one that can support life. This consideration seriously undermines the strong form of the anthropic principle, unless some argument can be used to rule out the possibility of other universes.

The physicist Lee Smolin of the Perimeter Institute for Theoretical Physics in Waterloo, Ontario, has proposed a different kind of multiverse. When certain very heavy stars run out of fuel, they collapse under their own gravity. If they are massive enough, then they collapse to a black hole, which is generally thought to be a point or nearly a point. Black holes are completely cut off from our universe in the sense that any matter or light that falls into one can never be retrieved. Smolin has noted that our universe began as a point or nearly a point and suggested that it originated as a black hole that was cut off from some other universe. He speculates that each black hole that is cut off from our universe may represent the beginning of another universe. If that is so, then Smolin's multiverse consists of a growing number of universes that are wholly inaccessible to each other, even in principle. Smolin goes on to suggest that some universes may be more adept at producing black holes than others; with time, they should be the most common. He further proposes a calculation roughly similar to Stenger's, in which the fundamental constants are varied to see which values produce the most black holes and therefore, presumably, the most daughter universes. If our universe turns out to have values close to the calculated values, then Smolin's theory will be supported. It will not, however, shed light on why those fundamental constants allow the development of life, nor whether only those fundamental constants allow the development of life.

## Conclusion

The strong form of the anthropic principle argues that the fundamental constants have been finely tuned to make the universe hospitable to life. Most of the universe, however, is probably inhospitable to life and is certainly inhospitable to life as we know it. Indeed, we know of only one place where the universe supports any life whatsoever, so it is hard to argue that the universe itself has been created for life. In addition, Stenger's toy universes cast serious doubt on the claim that the values of the fundamental constants in our universe are more conducive to life than other constants might be. Supporters of a special-earth theory thus argue that the earth has been specially positioned in the cosmos so that it will support life. If, however, there are billions or trillions of stars with planets, it is easy to see how conditions hospitable to life could arise somewhere purely by chance. Indeed, we could argue that life almost has to happen somewhere—and may well have happened more than once. Additionally, the special-earth hypothesis is untestable because it draws the same conclusion whether the universe as a whole is either hospitable to life or inhospitable. Finally, we point to suggestions that our universe is but one in a multiverse; if that is so, then it is easy to argue that we just happen to live in that universe where the conditions conducive to life pertain.

In this part, we have largely departed from our examination of biological evolution and turned to the physical sciences, mostly physics and geology. In Part V, we return to biology and palaeontology and discuss how our moral and religious sensibilities may well have deep biological roots.

THOUGHT QUESTIONS

1. List some events (in the sense we used the term in chapter 8) that appear to be very improbable, yet have occurred. Discuss them. If they were so improbable, why did they occur? Distinguish between improbable and impossible. Distinguish between prior probability and posterior probability.
2. How do we know the conditions necessary for the origin of life?
3. How do we know the conditions necessary for the evolution of intelligence?

PART FIVE

# *Evolution, Ethics, and Religion*

## Chapter 17

# *Evolution and Ethics*

I will jump into the river to save two brothers or eight cousins.

—J.B.S. Haldane, University College London, cofounder of the theory of population genetics

Evolution is amoral. The wasp parasitizes the caterpillar, the cheetah preys on the gazelle, and the bacterium infects the child, each without regard to its responsibility to the victim. Indeed, it has no responsibility to its victim. Nature, as Alfred, Lord Tennyson said, is "red in tooth and claw." Survival of the fittest. It's a dog-eat-dog world.

That, at least, was once the prevailing view. Humans, unlike other animals and bacteria, are thought to behave uniquely ethically or morally, even altruistically. If evolution or natural selection is amoral, then where did morality come from? One of the important advances in evolutionary biology is an understanding of how cooperation and even altruism may have evolved; evolution is no longer viewed as strictly competitive. At least some aspects of morality have very probably evolved by natural selection.

### Universal Code of Ethics

Many people think that, without God (or some other supernatural entity), there can be no morality beyond personal preference. Specifically, without God there can be no universal code of morality. Postulating that God gave us a fixed and universal moral code, however, opens serious philosophical and theological questions. Is something moral because God says it is, or does God say something is moral because it is moral? That is, did God choose arbitrarily, or did he have no choice? If God had chosen arbitrarily, then he could just as well have said that murder was ethical; on the other hand, if God had no choice, then why did he have no choice? Has the moral code more authority than God?

Rather than a single, universal code of morality, we find throughout the world and its cultures many codes that are sometimes similar or overlapping, sometimes different or nonoverlapping. We also find that moral codes may not be fixed in time. For example, the code of Hammurabi included a provision now known as *lex talionis*, or the law of retaliation. *Lex talionis* is the well-known law, "an eye for an eye, a tooth for a tooth," expressed in Exodus 21:23. *Lex talionis* may have originally been intended to preclude excessive punishment or blood feuds; it meant an eye for an eye literally but may have had the connotation, no excessive punishment, *no more than* an eye for an eye. Subsequent passages in Exodus suggest that the ancient Hebrews interpreted *lex talionis* as *the value of* an eye for an eye, but this interpretation may have been developed much later. Today, we are repelled by the idea of maiming someone in retaliation for a physical injury. What was considered moral at one time may sometimes be considered immoral at another time, though some moral precepts appear to be universal or nearly so.

Indeed, in early societies, the moral code extended primarily if not uniquely to members of the group. The Bible gives many examples in which the Hebrews—the *in-group*—wipe out or enslave other tribes—the *out-groups*—with the blessing of God. Columbus, on his very first voyage, kidnapped Indians and brought them to Spain as slaves. Our own society suspends the prohibition against murder during war by redefining it to exclude the killing of soldiers or, sometimes, even civilians, who are dehumanized as "collateral damage." It suspends the prohibition against slavery by importing merchandise made by workers who are little better than slaves. In practice, the moral code is not universal but often extends not far beyond members of our own group or nation. This consideration adds force to the argument that morality is somehow related to selection.

Saying that God gave us a universal moral code is the philosophical equivalent of intelligent-design creationism. It has no explanatory power and gives rise to a series of unanswerable and unenlightening questions. We need to look elsewhere for the origin of our moral or ethical code.

### The Evolution of Altruism

No less a scientist than Francis Collins, the director of the Human Genome Project, thinks that a Moral Law (his capitalization) exists outside humanity and suggests that it was given by God. He argues that selfless altruism cannot be accounted for by evolutionary principles and is a major problem for those who think it has a natural cause. Thus, Collins uses the existence of the moral law as an argument in favor of his own theism, though he avers that he might have to rethink that conclusion if evolutionary theory could ever satisfactorily account for the existence of altruism. His contention would be more convincing if humans were uniformly altruistic; rather, we find a spectrum of behaviors ranging from selfish to altruistic.

Nevertheless, it is fair to ask, "How can altruism have evolved?" First, we find cooperation, even altruism, at all levels of the animal kingdom. Indeed, certain single-celled amoebas, incorrectly called slime molds (*Dictyostelium discoideum*), exhibit behavior that can be considered analogous to altruistic behavior: a fraction of the colony is routinely sacrificed so that the colony can survive, though there is nothing deliberate about their behavior. During normal times, these amoebas are solitary and feed on bacteria. When food or water run scarce, however, the amoebas secrete a chemical known as cAMP; other amoebas are attracted toward higher concentrations of cAMP and also amplify the process by secreting additional cAMP themselves. Eventually, the amoebas are drawn together and form a clump known as a *slug*. The amoebas at the head of the slug form a stalk made primarily of cellulose and die in the process. Other amoebas climb the stalk and form a *fruiting body* that releases the spores from which the colony will reproduce when conditions are better.

Most of the amoebas in a given colony are clones of each other; that is, they have identical genes, because they reproduce asexually. We may understand their apparently altruistic behavior if we postulate that one of the mechanisms of evolution is the survival of the fittest *gene*, rather than the fittest *individual*. In this sense, genes fight to be passed to the next generation; to a gene it matters little which individual passes that gene to the next generation. When researchers Joan Strassmann and David Queller of Rice University mixed two clones in the laboratory, however, they found that the members of each clone struggled to stay away from the stalk and also that one clone was overrepresented in the fruiting body and therefore produced most of the spores. The amoebas, evidently, are altruistic only toward amoebas that share all their genes, that is, toward members of the in-group.

In addition, Strassmann and Queller identified a gene that allows the amoebas to adhere to one another and form the stem. They knocked that gene out of one strain of amoeba and introduced the resulting *knockout amoebas* into an otherwise genetically identical population. Unable to adhere, the knockout amoebas did not cooperate in the formation of the stalk but rather were overrepresented in the fruiting body. The researchers had discovered a single gene that, in a sense, coded for altruism. When they knocked it out, the knockout amoebas "cheated" and formed the majority of the spores. Such amoebas cannot survive in the wild because they cannot aggregate on their own, but they show how cheaters can prosper, at least for a time. Bear these amoebas in mind as you read further.

We are all familiar with the mother bird risking her life for her chicks, or with ants or bees sacrificing themselves for the queen and the colony. Mother birds that do not risk their lives are more likely to lose their chicks and not propagate their genes. Figuratively speaking, the mother bird risks her life because she "knows"—or her genes "know"—that without chicks, her genes will not propagate themselves. The mother bird that is driven to protect her genes more than herself is evidently more likely to pass her genes to succeeding generations than the mother bird that abandons her young to save herself. Why? In part because that mother bird may well not survive another year to reproduce again. Additionally, because the chicks are younger and more numerous, they may have greater overall potential for reproduction than the mother herself. Thus, to the genes, saving the offspring is worth the risk to the mother, so selection favors mothers that take that risk. In the same way, we might ask why all female ants do not compete to become the queen: because in most anthills there is a single queen, and ants have an unusual genetic makeup in which the females share three-quarters of their genes with the queen. It therefore makes no difference to the genes if many ants die defending the anthill, as long as the queen survives.

The primatologist Frans de Waal of the Yerkes Regional Primate Research Center in Atlanta has spent his career observing troops of chimpanzees in captivity. He thinks he has observed rudiments of morality among chimpanzees. In his book *Good Natured*, de Waal notes that gathering food is chancy. Perhaps one day one animal, as de Waal says, brings home the bacon, and another comes home empty-handed. On another day the second animal is successful, and the first unsuccessful. If they share, no one goes hungry after a bad day foraging, and the food supply becomes more reliable for both of them. De Waal relates an anecdote about two chimpanzees, one of whom generally shared food, whereas the other did not. When the generous chimpanzee needed food herself, others shared with her. By contrast, when the ungenerous chimpanzee needed food, others were unwilling to share with her. In all, de Waal and his team observed 5,000 interactions in which one chimpanzee had the opportunity to share food with another. As he had expected, chimpanzees willingly transferred food primarily to those other chimpanzees that had transferred food to them and not to chimpanzees that had not transferred food.

Chimpanzees and other animals groom each other or come to the aid of other nonrelatives without the prospect of an immediate reward. This practice is sometimes called *reciprocal altruism* and differs from ordinary cooperation because it has a delayed payoff. De Waal argues that cooperation can lead to reciprocal altruism. You have to remember who helped you, and you have to project who will help you in the future. Hence, we find reciprocal altruism primarily if not exclusively among the more-intelligent species. Michael Shermer, in his book *The Science of Good and Evil*, calls cooperative behavior among nonhuman animals *premoral behavior*. He thinks that morality is exclusively human, because morality requires an understanding of right and wrong, and nonhuman animals presumably do not consciously understand that distinction. De Waal, however, provides examples of chimpanzees comforting one another after an attack or sharing food with a hungry member of the troop. De Waal gives a striking example in which the alpha male excessively punished an adolescent male for copulating with one of the alpha male's favorite females. The females in the troop reacted indignantly and forced the alpha male to back down. De Waal suggests that chimpanzees understand right and wrong behavior.

De Waal thinks that apes, monkeys, and possibly elephants and whales are keenly aware of the needs of members of their own species and are concerned for them. He does not discuss how these animals acquired this concern but notes that human morality may have evolved from similar attributes. Evolution, suspects de Waal, provided humans with the prerequisites for morality, and human and animal societies provided the need. De Waal speculates that we are born with a moral ability that is analogous to our language ability, and we figure out our native moral system analogously to the way we figure out our native language.

Much of what we call altruism in humans can probably be explained in much the same manner as the altruism of the mother bird who defends her chicks. Humans, for example, are more likely to sacrifice for their children than for their nephews, more likely to sacrifice for their nephews than for their close associates, and more likely to sacrifice for their close associates than for outsiders.

Altruism toward relatives may be ascribed to *kin selection*. Specifically, natural selection favors organisms that are altruistic toward their relatives; your genes have a better chance of survival if you favor your relatives over complete strangers, because your relatives have many replicas of your genes. In the same way, in the tribal societies of our distant ancestors, it may have been beneficial to sacrifice for a close associate such as another member of your tribe. Indeed, in small bands of hunter-gatherers, much of the population may in fact have been related, though not so closely as cousins or second cousins. It is thus easy to see how humans and other tribal animals could have evolved altruism toward their relatives and members of their tribes, that is, toward the in-group.

Altruism toward your relatives may thus be at least partly evolved behavior. As humans joined larger and larger societies, sociological factors became more important, and we extended the scope of the in-group from the family to the tribe to the nation, and perhaps to all of humanity, although with successively diminishing strength. Humans can understand altruism and can extrapolate it beyond those in their immediate environment; they can easily apply the evolved tendency toward altruism to a widening circle outside the tribe and engage in apparently selfless acts of altruism, beyond merely favoring their kin and their associates.

Importantly, we are not arguing that all altruism has a biological explanation, and we do not claim that all human behavior is biologically determined. We think that both nature and nurture are important. We suggest, however, that altruism is not limited to humans and, in humans, must have a strong biological component. Indeed, it was just this biological component that led

Haldane to quip that he would risk sacrificing himself only for two brothers or eight cousins: each brother carries one-half of Haldane's genes, and each cousin, one-eighth.

Empathy is feeling what others feel; sympathy (the higher emotion) is being truly affected by what they feel. If you have sympathy for others, then you may develop true altruism, that is, not reciprocal altruism, but altruism with no expected payoff whatsoever. Cooperation combines with memory to give rise to reciprocal altruism, but reciprocal altruism combines with sympathy to give rise to true altruism, that is, altruism with no expectation of a reward.

Far from being built in, however, morality or altruism may vary with the environment or even in proportion to wealth or well-being. That is, the building blocks of morality can interact with the environment to produce behavior that we, in our environment, would consider immoral. For example, the anthropologist Colin Turnbull reports an African tribe that had apparently been dehumanized by extreme hunger. Turnbull never saw anyone over the age of three being fed by anyone else, and he saw people taking food away from their homes and eating surreptitiously so that they would not have to share the food. People evidently felt little or no sympathy for one another, and some literally stole food out of the mouths of weaker people.

### THE EVOLUTION OF COOPERATION

But if it's a dog-eat-dog world and every organism competes with every other organism for reproductive success, then where does cooperation come from in the first place? Theorists are beginning to home in on the origin of cooperation and find that cooperation sometimes supplements competition.

Let us begin with a game that is known as the *prisoner's dilemma*. Two men are charged with a crime and held in separate cells. If both prisoners remain silent, or *cooperate* (with each other), then they know that there will be no evidence, and both will go free. But each is given an offer: Confess and implicate your partner, or *defect*, and we will not only set you free but will also offer you a reward. Your partner will not only go to jail but also will be assessed a heavy fine to pay for your reward. It sounds like an offer they can't refuse, but there is a catch: If both prisoners defect, they are told, then there will be evidence aplenty, and they will both go to jail, but neither will pay a fine, because there is no reward. Neither prisoner can trust the other prisoner to cooperate, so each one minimizes his losses by defecting. Each goes to jail, but neither pays a fine.

Real-world decisions are rarely as stark as the prisoner's dilemma. Nevertheless, it is critically important to protect yourself against being betrayed. An animal that is too trusting may get eaten; a person who is too trusting (or too altruistic) may get taken for a sucker or, worse, get robbed. Why, then, do unrelated humans or animals ever cooperate with one another?

We may find part of the answer by imagining that we get to play the prisoner's dilemma over and over with the same person. After a while, we will acquire some data, and we will shortly learn whether to trust that person. A game in which the prisoner's dilemma is repeated many times is therefore more like real interactions between people.

In the 1970s, the political scientist Robert Axelrod organized a now-famous tournament in which computer programs played 200 consecutive rounds of prisoner's dilemma with each other. In each round, two competing programs were offered the choices given to the two prisoners. After each round, a program was awarded points depending on the outcome of that round. For example, if one program defected, but the other cooperated, then the cooperator got 5 points, and the defector got 0 in that round. The winning strategy was the simplest: a program called Tit for Tat, submitted by the game theorist Anatol Rapoport. Tit for Tat invariably begins the first round by cooperating with its opponent (Nice Tit For Tat); on subsequent rounds, it duplicates the opponent's previous move. It never defects first, but after the first round it is tough: If you cooperate with it, it cooperates with you; if you defect, it punishes you in the next round. Later research showed that Tit for Tat is not optimal under all conditions; for example, when there are many cooperators, it may be better to continue a winning strategy or to change a losing strategy without regard to the opponent's last move.

Psychologists and game theorists have studied another game called the *ultimatum game*. Here, one person, the *proposer*, is given a sum of money, say, $100. The proposer offers a second person, the *responder*, a share of the money. The responder may take the offer or leave it, but they may not negotiate. If the responder rejects the offer, neither the proposer nor the responder gets any money. The most rational strategy for the proposer is selfish: offer the responder a small fraction of the money, say, $10. Most proposers, however, offer a roughly equal share of the money. When they do not, the most rational strategy for the responder is to accept the money, no matter how little it is. In laboratory tests, however, most responders reject offers that they consider inequitable, and neither player gets any reward. Evidently, a sense of fairness defeats reason.

Martin Nowak and his colleagues at Harvard University performed a computer simulation of a repeated ultimatum game in which there were many players, and players knew the earlier behavior of other players. In other words, players got a reputation. The winning strategy then is to be fair, because other players will punish selfish proposers, whether or not they have played those proposers before. Fairness *becomes* rational. Nowak and his colleagues also allowed their players to reproduce and built mutation and natural selection into their model. Eventually, players adopted an equitable strategy.

How do games such as these work in the field? Imagine that a few tough cooperators arise by chance somewhere. They use Tit for Tat or some other strategy in their encounters with other individuals. When they meet each other, they cooperate and outperform their more-selfish rivals, who rarely cooperate. If they pass their cooperative nature on to their descendants, either biologically or culturally, then these tough cooperators will eventually prevail. Remember, evolution is a gradual process; the initial cooperators do not have to be particularly tough or particularly cooperative, as long as cooperating makes them slightly more fit than the other animals. Eventually, if the animals have good memory, then the cooperators will remember who generally cooperates with them and will not only cooperate in the short term but also will engage in reciprocal altruism, that is, cooperative behavior with a delayed payoff.

The evolutionary biologist David Sloan Wilson of the University of Binghamton has a somewhat different approach to the evolution of cooperation or altruism. Wilson has recently revived the controversial topic of *group selection* and applied it to cooperation. He considers a single, isolated group that is composed of *cooperators* and *shirkers*. The cooperators produce all the goods at some cost to themselves; the shirkers produce few goods at no cost to themselves. Selection favors the shirkers, and soon there are very few cooperators. Wilson's view is to some extent at odds with the preceding discussion, but let us follow it through.

Now, says Wilson, let us imagine several groups that are in competition. Each group contains cooperators and shirkers. Because the shirkers are relatively unproductive, groups with many shirkers are less fit than groups with few shirkers. Selection now favors not the cooperators themselves, but rather the groups that have a preponderance of cooperators.

Kin selection made people more likely to sacrifice for relatives than for complete strangers, because animals that are related share many genes. If groups compete, it may be advantageous for people to sacrifice for other members of the group, even if those members are not biologically related. Thus, a group may outcompete other groups because some members will sacrifice for the good of the group or will cooperate within the group.

A recent computer simulation pitted virtual groups against one another and found that within-group cooperators survived as a result of between-group competition. The experimenters, Jung-Kyoo Choi of Kyungpook National University, Korea, and Samuel Bowles of the Santa Fe Institute, simulated groups of "organisms" with two genes. One of these genes related to within-group behavior and had one allele for cooperation and one for shirking; the other related to outside-group behavior and had one allele for trading with foreigners and one for parochialism. Choi and Bowles argue that their simulation reflects the realities of our species in the Pleistocene era, when groups would sometimes have competed fiercely for resources. When Choi and Bowles pitted virtual groups against one another, they found that the most successful groups had a preponderance of either parochial cooperators or trading shirkers.

Not surprisingly, the least-successful groups had a preponderance of cooperative fighters; such organisms sacrifice themselves in war as well as pay a price for being altruistic within their own groups. More surprisingly, however, these groups, while generally unsuccessful, did not wholly disappear and included a proportion of altruists. Thus, it seems that group selection can produce both altruists and hostile xenophobes, even though individual selection discriminates against both. This result is also at odds with the earlier discussion. Clearly more work needs to be done, but surely we may not dismiss out of hand the possibility that altruism can evolve by natural or group selection.

## Conclusion

In this chapter, we have shown how cooperation is found at all levels of the animal kingdom and has evolved by natural selection. In more-intelligent animals, cooperation can evolve into reciprocal altruism. Animals, such as humans, that feel sympathy for others can further evolve from reciprocal altruism to true altruism. Even so, we find that people are more cooperative with and more altruistic toward their relatives and members of their group, precisely as we would expect if cooperation and altruism were the result of natural selection.

In the next chapter we will show why we think that science and religion are not incompatible, though, frankly, some religious beliefs are incompatible with scientific knowledge. We take issue with Stephen Jay Gould's "nonoverlapping magisteria," however, and argue that rational religious belief must of necessity take into account all that science has to offer. We conclude the chapter by examining the acceptance of evolution by many religious denominations within the United States.

## Thought Questions

1. Suppose that a moral code is in fact a human or societal construct that has its basis in evolutionary biology. Would that fact in any way alter your behavior? Do people in general need the threat of punishment to behave ethically? Do you? Explain your answers.
2. Give examples of altruism you have encountered in your own life. Try to explain them purely biologically using this chapter as a guide.
3. If an eye can evolve from an eyespot, can a moral ability or a moral code evolve from reciprocal altruism or from premoral behavior? Why or why not?

Chapter 18

# *Why Science and Religion Are Compatible*

. . . the anticreationist cause in the U.S. would be doomed without the help of Christians who are favorably inclined toward the teaching of evolution.

—Frederick Crews, University of California, Berkeley, prominent literary critic

Science and religion have more in common than many people think. According to the philosopher of religion Ian Barbour, both formulate hypotheses, both test their hypotheses, and both include an element of faith. Science, however, requires precise hypotheses, tested rigorously against empirical standards; if it relies on faith, it is a very different kind of faith from religious faith.

In his book *The Radical Reform of Christianity: A Focus on Catholicism*, religious studies professor Edward Brennan of Cleveland State University argues that "both science and spiritual teaching are entirely problem-based." But, Brennan explains, religious extremists commonly err when they think that spiritual teaching is dogma, "as if it is there to tell us what to do." Many people do not recognize that religion formulates and tests hypotheses. Even a religion that stresses faith above all else seeks evidence to support its faith. That is why we hear so many testimonials to the fact that someone "got religion" and turned his or her life around or was cured of a serious disease. These testimonials are a form of hypothesis testing, though they may not be clearly understood as such. Unfortunately, they are usually irreproducible anecdotes, and they are often cherry-picked to support a preconceived notion. Quantitative efforts to demonstrate statistically that prayer can affect the outcome of a disease have been equivocal at best.

Scientists, on the other hand, often display faith in their theories. Quantum mechanics is one of the most successful theories ever developed. Many physicists interpret quantum mechanics as predicting that nature is wholly random at the subatomic level. Though he did not deny the validity of quantum mechanics, Einstein never accepted that interpretation, in part because of his faith that the universe could be understood rationally. Pauli, the discoverer of the neutrino (see chapter 5), likewise had faith in the law of conservation of energy and, as a direct result, predicted a new particle. Good scientists, however, do not carry faith too far and, like Deryagin, the discoverer of polywater, routinely admit error or abandon a cherished theory when required by the evidence (see chapter 4). Religious believers, too, sometimes abandon a cherished belief, but it is usually substantially harder to convince them to do so. Indeed, in some religious traditions, believing in the absence of evidence is considered a virtue.

Some scientific theories are so firmly established that we may safely say they are correct, at least within certain limits. Newton's theory, for instance, is correct for weak or moderate gravitational fields and low velocities (but it has been superseded by Einstein's theory of relativity in strong fields and high velocities). Biological evolution or, more properly, descent with modification, is so well established that we may consider it a fact. The modern theory of evolution explains that fact better than any other theory and is one of the most well-established theories in science—on a par with Newton's theory of gravity or Einstein's theory of relativity.

Are science and religion compatible? That is the wrong question; the legions of religious scientists show that these fields are in general compatible in the sense that thoughtful people can be both scientists and believers. The correct question is, "Are science and certain religious beliefs compatible?" The answer is clearly "no": Certain religious beliefs are not compatible with certain scientific facts.

### Up from Literalism

Stephen Godfrey and Christopher Smith both began their careers as young-earth creationists. Godfrey, however, became a palaeontologist, and Smith, a Baptist minister. Each underwent what they call a "pilgrimage," as the acquisition of compelling, new knowledge forced them to reevaluate their literalist religious beliefs. Both came to accept both biological evolution and the antiquity of the universe, yet both remained devout Christians.

In the 1980s, Godfrey enrolled in graduate school, where he studied vertebrate palaeontology. One of his first jobs was to search for fossils in sedimentary rocks. These rocks are layered, so the deeper you dig, the older are the fossils you find (see chapter 15). Godfrey was most impressed by fossilized footprints and other markings, known as *trace fossils*, left in the sandstone by prehistoric animals. As a young-earth creationist, Godfrey thought that the sedimentary rocks and the fossils within them had been laid down by the Noachian flood. If that was so, then how could terrestrial vertebrates have left footprints in the sand (which was presumably under water)? Godfrey researched trace fossils and found that they appear in many different strata in many sedimentary rock formations all around the world. He could not account for the appearance of trace fossils in rocks that had supposedly been left behind by a flood that had killed the very animals that might have made the footprints. Godfrey also found cracked and fossilized mud flats, which he recognized immediately had been baked by the sun and could not have been deposited by a flood. Earth suddenly became much older than Godfrey had imagined.

Godfrey presents further evidence that convinced him that God had not created every species from scratch. Perhaps God had decided to use natural processes when creating species. Why not? The Bible, as Godfrey notes, says that God sends rain upon the face of the earth. Yet no one rejects the science of meteorology or argues that rain is not the result of evaporation and condensation. No one demands that "biblical meteorology" be given equal time in science classes. Considerations such as these have convinced Godfrey that evolution is no more antireligious than meteorology; both are equally naturalistic explanations of observed facts.

Smith is Godfrey's brother-in-law and collaborated with him on the book *Paradigms on Pilgrimage*. Smith's story, related in the second half of the book, is similar to Godfrey's, except that Smith became a creationist while in high school. Smith's equivalent of Godfrey's trace fossils was a course in Old Testament *hermeneutics* (interpretation of scripture), in which he was exposed to refutations of day-age theory and gap theory (see chapter 6). His professor

EFFICACY OF PRAYER

Does prayer work? To the extent that it makes someone feel better, the answer is obviously yes. Additionally, there is evidence that churchgoers are healthier than non-churchgoers. Such evidence is scientifically weak, however, and does not establish causality—maybe people who are happy and healthy are drawn toward church attendance. To examine prayer scientifically, we need better evidence.

Medical researchers know that if they give patients a worthless treatment, known as a *placebo*, then sometimes up to 30 percent of those patients will appear to have been cured. To guard against the placebo effect, medical researchers employ a *double-blind test* and treat some patients with the medications they are testing and other patients with a placebo. It is important that neither the physician nor the patient knows who gets the medication and who gets the placebo. A medication is deemed efficacious only if statistics show that it is significantly more efficacious than a placebo.

It is difficult to test for the efficacy of prayer because of the placebo effect; if an individual prays for herself, she obviously knows she is being prayed for, so it is impossible to distinguish a real effect from the placebo effect. In order to perform a scientific evaluation of the efficacy of prayer, we have to devise a double-blind test in which some people are prayed for without their knowledge and some people, who serve as a *control group*, are not prayed for. Even double-blind tests are difficult, however, because of the probability that people extraneous to the experiment are praying for some of the patients. Still, we do not agree with some critics that these background prayers wholly invalidate the experiment, any more than background noise makes speech unintelligible.

Randolph C. Byrd of the San Francisco General Hospital devised a double-blind procedure to measure the efficacy of *intercessory* prayers, or prayers by third parties asking God to intervene on behalf of someone else. He divided approximately 400 cardiac patients into a treatment group and a placebo group, and compared the outcomes of the two groups. Byrd claimed a positive result for the treatment group, and his study was widely hailed as a landmark. His study was flawed, however, in that it was imperfectly blinded, and Byrd may have selected some of the criteria for evaluating the outcomes after the experiment was performed, not before. More importantly, though, as pointed out by Gary Posner, a physician specializing in internal medicine, the bottom line was that the treatment group spent the same time, on average, in the hospital and in the coronary care unit as the placebo group, and went home with the same number of medications.

Following Byrd, Herbert Benson of Harvard University, the physician who once tried to put meditation on a sound scientific basis, conducted a multiyear Study of the Therapeutic Effects of Intercessory Prayer, or STEP trial. Benson divided approximately 1,800 patients into three groups (not two): patients who were informed that they might receive intercessory prayer and received it; patients who were informed that they might receive intercessory prayer and did not receive it; and patients who were correctly informed that they would receive intercessory prayer. Two Catholic groups and one Protestant group were enlisted to pray for the patients (Benson, incidentally, is Jewish). The result of the STEP trial will not give comfort to those who believe in the efficacy of prayer: the outcomes were statistically identical for the first two groups, but the third group, the patients who knew that people were praying for them, had statistically poorer outcomes than the other two groups. Benson speculated, in effect, that they might have been the victims of a reverse-placebo effect. Another study, called MANTRA II, likewise showed no benefit from intercessory prayers.

further introduced him to the idea that the opening chapters of Genesis are poetry, because they include purposeful repetition of vowel and consonant sounds (alliteration and assonance), and display the "rhyming thoughts" characteristic of Hebrew poetry. To Smith, Genesis does not describe how we got here but rather why we got here. Its purpose is to instruct us to follow God's leadership; it is not a history or a *cosmogony* (a theory of the origin of the universe). Consequently, because the first chapters of Genesis are poetry and not cosmogony, they do not preclude a very ancient universe.

For Smith, the biblical authors (note the plural noun) saw the earth as a flat disk under a solid, domed sky. Thus, the author of Ecclesiastes thought that streams run to the sea and then recycle the water endlessly, a statement that could barely be defended as a description of the hydrologic cycle. But the author also says that the sun rises and sets, and then hurries back to the place from which it rose, a statement that is simply not true. The biblical authors were writing what they saw or thought they saw, not what is objectively true. Why then, asks Smith, do we not interpret Genesis 1 in the same way? That is, why do we interpret Genesis 1 as a cosmogony and not as the poetic description it so plainly is?

The final chapter of *Paradigms* is written by both authors and is what they call a "close reading" of Genesis 1. The authors make a concerted effort to interpret the chapter as it would have been read or understood by the biblical authors and their readers. We will not go into detail, but rather we will concentrate on one example. On the first day, God created light, but he did not create the sun until the fourth day. This seeming inconsistency is a problem for anyone who thinks that the account in Genesis parallels modern cosmological thinking.

The authors of Genesis, however, were recording what they saw. As the poet Dylan Thomas might have said, they saw light break where no sun shone: at twilight and on cloudy days. Evidently, say Godfrey and Smith, the biblical authors did not associate the light of twilight with sunlight. They saw two regularities and recorded them: First dawn came, then the sun came up. Hence, it was no contradiction that God created light on the first day and the sun on the fourth day. We see a contradiction only if we read the Bible anachronistically and think that the book of Genesis is a literal history. Godfrey and Smith use this kind of reasoning to show that not every statement in the Bible is factually correct, especially if it is erroneously subjected to a modern interpretation.

What we are about to say is delicate, but, frankly, science tests its hypotheses better than any other human endeavor. Scientists do their best to test (even falsify) their own hypotheses before they will accept them. Any belief, religious or other, that denies known scientific fact is seriously in need of reconsideration. Religion and science are not incompatible, but some religious beliefs are at odds with facts and need to be reevaluated. Unhappily, rather than reevaluate their beliefs, proponents of such religious beliefs have set forward the pseudo-scientific claims that are a major concern of this book.

### Overlapping Magisteria

A *magisterium* is the authority to teach a subject; the term derives from the Latin word *magister*, teacher or master. Usually, it refers to the authority of the Roman Catholic Church, but the distinguished paleontologist Stephen Jay Gould borrowed the term and used it to mean the authority to teach any given subject. Thus, to Gould, science has its magisterium—facts, data, and theory—and religion has its—ultimate meaning and morality. These magisteria, Gould says, are nonoverlapping: science does not discuss ultimate meaning or morality, and religion does not discuss empirical facts or scientific theories. Gould called his principle NOMA, for nonoverlapping magisteria. As Galileo is supposed to have said, "The Bible tells us how to go to heaven, not how the heavens go."

Gould thus asks for a truce between science and religion: You stay out of my magisterium, and I will stay out of yours. Unfortunately, matters are not so simple. As we saw in chapter 16, science may have much to say about the evolution of morality, yet morality is a subject that Gould would leave to religion. Similarly, many churches (and not only fundamentalist churches) have dogmas that lead them to contradict at least certain findings of science. Gould's NOMA principle would require these churches, in effect, to throw in the towel and concede certain points that they may not be willing to concede. The National Academy of Sciences subscribes to some degree to the NOMA principle.

The Roman Catholic Church is the largest religious denomination in the world, with over 1.1 billion adherents, and its intellectual leadership is among the most influential. What does the Catholic Church have to say about evolution? The first Vatican Council was held approximately ten years after the publication of *On the Origin of Species*. The report of the council does not mention evolution specifically, but it states that "faith is above reason" and Christians are forbidden to defend scientific conclusions that are "known to be contrary to the doctrine of the faith." It goes on to assert that "faith delivers reason from errors." In 1909, however, the Church decreed that the concept of *special creation*, or the creation of species, applied only to human beings, not to other species. In 1950, Pope Pius XII issued an encyclical, *Humani Generis*, and ruled that the study of evolution was not forbidden. Catholics, he said, may form their own opinions but should not confuse fact with conjecture. They must, however, believe that the human soul was created by God, that is, did not derive from inanimate matter. Finally, all humans have descended from a single individual, Adam, and not from a small group of men and women (see chapter 1).

In 1996, Pope John Paul II addressed the Pontifical Academy of Sciences and recognized that the theory of evolution had very significant support. He went on, however, to stress that the human soul was created directly by God and further that the human mind did not emerge as a result of natural processes. More recent writings by Cardinals Joseph Ratzinger (now Pope Benedict XVI) and Christoph Schönborn suggest that the Church is now leaning toward a position close to intelligent-design creationism, but it accepts that chance could be consistent with God's plan. The Church is thus claiming a middle ground between philosophical naturalism (see chapter 3) and creationism.

Most mainline Protestant churches in the United States accept a nonliteralist reading of the Bible and espouse views that are not inconsistent with evolution. In approximately decreasing order of membership, these churches include the United Methodist Church, the Evangelical Lutheran Church in America, the African Methodist Episcopal Church, Latter-Day Saints (Mormons), the Episcopalian Church, many Reformed Churches such as the Presbyterian Church (USA), Orthodox Churches, American Baptist Churches, and the United Church of Christ. The Unitarian-Universalist Church and most Jewish denominations likewise hold views that do not rule out an acceptance of evolution. Data on U.S. Muslims are hard to find, but a majority is foreign born, and many are upper class and secular; there is no reason to think that they are significantly less likely to accept evolution than the rest of the population. The total U.S. membership of these churches and religious organizations, including the Roman Catholic Church, is on the order of 100 million.

Other churches require a belief in a literal interpretation of the Bible and hold that evolution is inconsistent with this belief. Such churches include the Southern Baptist Convention and other conservative Baptist churches, Assemblies of God and other Pentecostal churches, the Missouri and Wisconsin synods of the Lutheran Church, the Seventh-Day Adventist Church, Jehovah's Witnesses, and the Christian Science Church. In addition, some ultraorthodox Jews hold to a literal interpretation of the Bible. The total U.S. membership of these churches and religious organizations is over 40 million.

The overall church membership in the United States, including non-Christian denominations, is over 150 million. Only one-fourth to one-third of that membership belongs to a denomination that formally precludes acceptance of evolution. The public as a whole is not so accepting, as we see from polling data.

Polling data have given consistent results over the years, as long as the questions asked are consistent. A series of seven Gallup polls taken between 1982 and 2006 found that 11 percent of Americans thought, "Human beings have developed over millions of years from less advanced forms of life, but God had no part in this process"; 38 percent thought that "God guided this process"; and 45 percent thought that "God created human beings pretty much in their present form at one time within the last 10,000 years or so." The *margin of error* of such polls is typically about 3 percent. That is, roughly, if the poll were repeated with a different sample (and if the sample is truly representative of the population), the probability is 95 percent that the result would be the same, within 3 percent either way. In fact, during the twenty-five years of polling, no result differed from the averages given above by more than 3 percent. Thus, it seems likely that nearly half of the population of the United States are young-earth creationists or are at least sympathetic to that position.

Other data support this conclusion. Polls by *Newsweek*, CBS News, and CNN/*USA Today*/Gallup revealed that 48 to 53 percent of the public thought that God had created humans in their present form either within the last 10,000 years or so, or exactly as the Bible describes. When CBS News omitted the restriction to the last 10,000 years, the percentage increased from 46 percent to 53 percent. Other polls have found that 50 percent to 66 percent thought that creationism was definitely or probably true.

Additionally, a poll by Pew Research Center for the People and the Press, and the Pew Forum on Religion and Public Life revealed that 58 percent favored teaching creationism alongside evolution in the public schools, whereas only 12 percent favored teaching evolution only. Likewise, a CBS News/*New York Times* poll similarly found that 65 percent favored teaching creationism. A Harris poll found that 55 percent favored teaching evolution, creationism, and intelligent-design creationism in the public schools. Reacting to the Pew poll, Eugenie C. Scott, the director of the National Center for Science Education, suggested to the *New York Times* that people thought it was fair to give creationism a hearing: "Americans react very positively to the fairness or equal time kind of argument. In fact, it's the strongest thing that creationists have got going for them because their science is dismal."

### Conclusion

We distinguish between rational, irrational, and nonrational (see box). A religious belief that is consistent with known scientific facts may be *non*-rational, but not necessarily *ir*rational. Smith and Godfrey are paradigms of people who have reevaluated, but not necessarily disowned, their religious beliefs precisely because those beliefs came into conflict with what they knew to be scientific fact (in Godfrey's case) or with a cogent understanding of the Bible (in Smith's case). That is not to say that scientific fact and religious belief can be isolated from each other, nor that they can always be harmonized. Stephen Jay Gould, however, was seriously mistaken in thinking that science and religion are separate magisteria and have nothing to say to each other. To the contrary, science must set limits on what it is possible to believe, and religion, though it does not have a monopoly on morality, has much to say to science as well.

### RATIONAL, IRRATIONAL, AND NONRATIONAL

A *rational* belief is a belief that is based on reason and supported by facts, evidence, and logic, or some combination of those. Not all beliefs can be supported by facts, evidence, or logic, yet they are not necessarily *irrational* beliefs, that is, beliefs that are contrary to reason or not in accord with facts. In his poem, "Ode on a Grecian Urn," John Keats wrote, "A thing of beauty is a joy forever." But if you do not like Grecian urns, there is no rational argument, no fact that can be adduced to convince you that they are indeed things of beauty. Similarly, the belief that all people are created equal cannot be supported by facts, yet it is not irrational to hold that belief. The more or less opposite belief—that people should be pawns of the state—likewise cannot be supported or refuted by facts, however distasteful we may find it. Yet we would not call such beliefs irrational, because they are not claims about facts and they do not fly in the face of evidence. Instead, we call these beliefs *nonrational*—not based on reason, yet not in conflict with reason.

None of these nonrational beliefs involves a claim of fact. Beliefs that involve claims of fact but cannot be conclusively supported or refuted rationally are (or should be) considered closer to hypotheses. Thus, a belief in God is a claim of fact and may be considered nonrational, provided that it is somewhat tentative and does not force the believer to accept a proposition that is contrary to known fact. Stephen Godfrey recognized that his religious belief had led him to accept propositions that are contrary to fact, and he wisely modified his belief accordingly.

The rationality or irrationality of a belief, incidentally, does not bear on the truth of that belief. A strident atheist—one who insists that there is no deity, as opposed to one who thinks that the weight of the evidence most likely militates against a deity—cannot support his or her claim of fact with hard data or cold logic and so is irrational in the sense that we use the term. The atheist's belief could be correct, but it is still irrational if held with certainty. Indeed, what is considered rational at one time may be irrational at another time. For example, the medieval philosopher Moses Maimonides argued that a man could never fly. At the time, that was a rational belief, yet today anyone promulgating that belief would justifiably be considered irrational. In the same way, before many of the discoveries of modern science, creationism was rational; today, as Godfrey and Christopher Smith have discovered, it leads only to beliefs that contradict factual evidence and is therefore irrational.

Thought Questions

1. The philosopher Bertrand Russell wrote that you could not disprove an assertion that there was a teapot orbiting the sun somewhere between the orbits of Earth and Mars, provided that you added that the teapot was too small to be observed with a telescope (space travel was not a reality in 1952 when Russell wrote his essay). A belief in Russell's "celestial teapot" is harmless and not in conflict with any known scientific fact. Is such a belief irrational or nonrational? Is Russell justified in comparing a belief in God with a belief in a celestial teapot? Explain your answers.
2. Can an evolutionary biologist simultaneously accept the Catholic Church's position that all humans descended from a single person, Adam, and the scientific position that there were many males at the time of Y-chromosomal Adam? Simultaneously holding two conflicting positions is sometimes called *compartmentalization*. Some people think that compartmentalization is necessarily irrational. Do you agree? Why?
3. We often hear people say things like, "If I wash my car, it will rain." Such reasoning is called *mythopoetic*. Give other examples of mythopoetic reasoning, including some that are serious. Do you think that religion could have evolved from mythopoetic thinking? What if it did?

CHAPTER 19

# *Summary and Conclusion*

Tell them what you are going to tell them; tell it to them; then tell them what you have just told them.

—Author unknown

Then summarize.

—Matt Young and Paul Strode

THIS BOOK HAS BEEN a brief arguing in favor of evolution in particular and science in general, and against all forms of creationism. We began by demonstrating why evolution is important and also why people are, frankly, afraid of it. We noted, in particular, that scientific facts are true whether or not we like them and, indeed, whether or not we understand them. Further, we consider it intellectually dishonest to deny a well-supported scientific theory such as evolution on aesthetic grounds or by appealing to an inappropriate authority. Descent with modification is a fact that is supported by evidence from palaeontology, phylogeny, genetics, and biogeography. The modern synthesis, a combination of Darwin's theory with Mendel's genetics and advanced mathematics, provides by far the best explanation for the observed fact of descent with modification.

We showed in chapter 3 that creationism is a product of a literal reading of the Bible. We claim that creationism is flatly wrong. We argue later, in chapter 18, that religion and science are not necessarily incompatible, even though creationism and science are wholly incompatible. Indeed, we think that no religious belief is necessarily incompatible with science, as long as it makes no claims of fact that are demonstrably false. In the same way, science cannot truly pronounce religious beliefs to be wrong, as long as those beliefs are not testable by the methods of science. A great many scientists, including evolutionary biologists, are themselves religious.

Young-earth creationists deny principles that lie at the very foundation of modern science: descent with modification, cosmology, geology, and radioactivity. Old-earth creationists distort both the Bible and modern cosmology in a vain effort to demonstrate that the Bible is true in the sense of writing actual history, but they do not deny the principles of science. Intelligent-design creationism is in some sense a branch of old-earth creationism. Intelligent-design creationists do not deny the antiquity of either the Earth or the universe, and they do not deny common descent. They recognize, for example, that humans have descended from an apelike ancestor. They claim, however, that evolution can go only so far without intervention. Their arguments are flawed in many ways, and, like young-earth creationists, they seem to think that they can discredit the theory of evolution if only they can find something it cannot explain. Further, both intelligent-design creationists and young-earth creationists presume that, if they succeed in discrediting the theory of evolution, then their own pet theory will win by default. Science, as we have seen, does not work that way. Theories do not win by default, and they do not lose because they are not complete; rather, theories win by explaining a preponderance of the evidence and explaining it better than competing theories. Creationists have explained nothing whatsoever. In the unlikely event that evolutionary theory was somehow overthrown, creationism would have a long way to go before it was accepted by the scientific community.

In short, creationism is more than a pseudoscience. Intelligent-design creationism, at least, is an attack on all of science, because it condemns the methodological naturalism that is at the heart of the scientific method. If the authors of the Wedge Document have their way, hard science will be replaced by an unspecified theistic understanding of science. We find it hard to understand how science can operate, for example, if God is assumed to step in every so often and provided a harmful parasite with resistance to a drug. Science must assume, with theistic evolution, that the world is regular and God does not step in periodically.

For these reasons, we have tried to show in this book not only how creationism is wanting and how pseudoscience works but also how evolutionary biology really works. We heartily recommend that anyone who criticizes evolutionary biology first learn about it.

# Glossary

**academic freedom.** The principle that teachers and professors must be free to pursue their research and teaching activities without external interference, as by politicians or administrators.

**accretion.** In astronomy, the increase of mass of a celestial object such as a planet as it collects surrounding objects as a result of its gravitational attraction.

**ad hoc hypothesis.** See *auxiliary hypothesis.*

**adaptation.** A genetically based behavior, structure, or function that arises in a population as a result of genetic change, is favored by natural selection, and increases the fitness of individuals in the population that possess the adaptation.

**aerobic respiration.** In an organism, the process of breaking down biological molecules in the presence of oxygen.

**agglomerated complexity.** A *meaningful message* that has been generated by combining two or more meaningful messages and whose information as a result exceeds 500 bits.

**agnosticism.** The belief that we cannot know whether or not God (or other deities) exists (see also *atheism*, *deism*, *theism*, *theistic evolutionism*).

**algorithm.** A step-by-step procedure for solving a mathematical problem; especially, a computational procedure carried out on a computer.

**allele.** One of two or more alternative forms of a *gene.*

**altruism.** In our context, providing a benefit to another without expectation of reciprocity (see also *reciprocal altruism*).

**amoeboid movement.** A creeping type of movement of individual cells in which the cytoplasm flows into projections of the cell membrane and the rest of the cell and its contents follow.

**anthropic principle.** 1. Weak anthropic principle: If the conditions were not right for our existence, then we would not exist. Also *weak form*. 2. Strong anthropic principle: Conditions are right for our existence precisely so that we will exist. Also *strong form* (see also *fine-tuning, fundamental constant*).

**appeal to authority.** An argument in which the truth of an assertion is based on the testimony of one or more presumed authorities. Though strictly a fallacy in logic, the appeal to authority lends weight to an assertion, provided that the authority is truly expert.

**appeal to the consequence.** A logical fallacy wherein a claim of fact is rejected because its consequences are deemed to be undesirable.

**archean.** Pertaining to the oldest known rocks.

**atavism.** The appearance in an organism of a trait or character that was known only in an ancestral form. Also called a *throwback*.

**atheism.** 1. The belief that the existence of God (or other deities) has not been proved and hence there is no reason to believe in them. 2. Sometimes, the firm belief that there is no God (or other deities) (see also *agnosticism, deism, theism, theistic evolutionism*).

**autotrophic.** Relating to an organism that can synthesize its own food from nonnutritive inorganic molecules (see also *heterotrophic*).

**auxiliary hypothesis.** A *hypothesis* that is employed to explain an apparently anomalous result. Also ad hoc hypothesis.

**bacteriophage.** A virus that can infect only bacteria.

**balanced treatment.** The requirement to teach creation science if evolution is taught.

**basal.** Situated at the base or foundation of an evolutionary tree.

**Batesian mimicry.** A form of *mimicry* in which the model has an attribute (poisonous, for example) that makes it unprofitable to a predator, but the mimic lacks the unprofitable attribute.

**biblical inerrancy.** The view that the original manuscript of the Bible was completely free of error (see also *biblical literalism*).

**biblical literalism.** The view that every apparently factual statement in the Bible is literally true and is not subject to interpretation (see also *biblical inerrancy*).

**big bang.** The cosmic explosion that denotes the beginning of our universe 13.7 billion years ago. Also, the cosmological theory that the universe began as a small, hot fireball that has expanded to its present state and continues to expand (see also *steady-state theory*).

**biogenic law.** The principle, formulated by Ernst Haeckel, that the developmental pathway of an embryo, from fertilized egg to adult, mirrors its evolutionary history. Contrast with *von Baer's law*.

**biped.** A two-legged animal.

**bit.** The unit of measure of information.

**black hole.** A region of space whose gravitational field is so strong that nothing, not even light, can escape.

**blackbody.** In physics, a perfect absorber or radiator. A blackbody emits a well-known *spectrum* of radiation that depends on its temperature.

**blind spot.** The region of the *retina* which is located a few millimeters on the nasal side of the *fovea* and in which there are no photoreceptors owing to the intrusion of the *optic nerve*. Also *optic disk*.

**blind test.** A test that is performed in such a way that the experimenter cannot see or influence the outcome (see also *double-blind test*, *experimenter bias*).

**camouflage.** A color or color pattern that makes it difficult for an individual organism to be distinguished from its nonliving surroundings (see also *mimicry*).

**carrier.** An individual that possesses a mutant or alternative form of a gene whose symptom is not expressed.

**cataract.** An opacity of the *lens* of the eye that results in impaired vision or blindness.

**catastrophism.** In geology, the theory that the earth has been affected by sudden, violent events that were sometimes worldwide (see also *uniformitarianism*).

**character.** In biology, a structural or functional attribute or trait of a group of organisms that is determined by a *gene* or group of genes and can be used to define the group and separate it from other groups.

**chemosynthesis.** In biology, the conversion of carbon and nutrients into organic matter using oxidation rather than light as an energy source.

**cilium.** Microscopic hairlike projection from a cell membrane that aids in locomotion or moving materials past the cell (see also *flagellum*).

**coevolution.** The evolution of two or more species as a direct result of their interaction, as coevolution between predator and prey.

**common ancestor.** A group of organisms from which several related groups of organisms evolved.

**compartmentalization.** Separating a concept into parts or compartments, often artificially and sometimes with the intention to oversimplify.

**complex specified information.** See *specified complexity.*

**concordia.** In geology, a theoretical curve used in *radiometric dating* with the uranium-lead system.

**cone.** The light-sensitive receptor that is found in the *retina* and provides high-acuity vision in bright light (see also *rod*).

**confirmation bias.** See *experimenter bias.*

**conjugation.** The transfer of genetic material between two individual bacterial cells.

**continental drift.** The motion of the Earth's continents with respect to one another.

**convergent evolution.** The evolution of similar adaptations in species that do not share a common ancestor (are not closely related) because the environmental pressures on them over evolutionary time are the same or very similar.

**cornea.** The transparent, convex portion of the eyeball that covers the *iris* and helps focus light onto the *retina.*

**cosmic background radiation.** See *cosmic microwave background.*

**cosmic microwave background.** Electromagnetic radiation that fills the entire universe and whose *spectrum* corresponds to a *blackbody* whose temperature is approximately 2.7 kelvins; such radiation is a relic of the *big bang.*

**cosmogony.** The study of the origin and evolution of the universe.

**creation science.** The attempt to use the scientific method to prove the account of the creation given in Genesis. Also called scientific creationism.

**creationism.** See *creation science, young-Earth creationism, old-Earth creationism, intelligent-design creationism.*

**day-age theory.** The view that the six days of creation represent ages rather than literal twenty-four-hour days (see also *gap theory, old-Earth creationism*).

**deism.** The view that God created the universe, left it to its own devices, and exerts no influence on natural phenomena (see also *agnosticism*, *atheism*, *theism*, *theistic evolutionism*).

**demarcation problem.** In philosophy of science, the problem of demarcating science from nonscience and from pseudoscience.

**descent with modification.** The process by which all living things, both past and present, have changed in structure and function from a *common ancestor* over many generations.

**development.** The process by which an individual organism changes in physical form through its lifetime, beginning with a fertilized egg and ending with the adult.

**differential reproductive success.** Differences in the average number of reproductively viable offspring among groups of individuals in a population that result from different heritable characteristics among those groups.

**diffusion.** 1. Spontaneous mixing of two or more substances. 2. Net motion of one substance, such as a dissolved compound, from a region of high concentration to a region of lower concentration.

**diffusion coefficient.** In physics or chemistry, a number that describes the rate of diffusion, as of one material into another, and depends on temperature and viscosity.

**digitigrade.** A form of movement by animals that walk on either two legs or four wherein the toe bones are the only parts of the foot that are in contact with the ground during each step (see also *plantigrade*, *unguligrade*).

**dipole moment.** In physics, a measure of the strength of a very small magnet that consists of one north pole and one south pole.

**discontinuous function.** In mathematics, a function for which small changes in the input result in very large changes in the output.

**discordia.** In geology, the line that connects the data points in *radiometric dating* with the uranium-lead system.

**display hypothesis.** The hypothesis that the testicles of many mammalian males move into the scrotum outside the body cavity during *development* because the presence of a brightly colored scrotum of many species indicates a male's readiness to mate (see also *inguinal hernia*, *temperature hypothesis*, *testicular descent*, *training hypothesis*).

**documentary hypothesis.** A hypothesis, now well established, that the five books of Moses were compiled by an unknown redactor from four independent sources (see also *higher criticism*).

**dogma.** In religion, a doctrine declared to be authoritative by a church or other religious organization.

**Doppler effect.** In physics, an apparent shift of the frequency of a wave when the source and the receiver are in relative motion. Specifically, the shift is toward lower frequency (longer wavelength) when the source and receiver recede from each other (red shift).

**double-blind test.** A *blind test* that uses human subjects who do not know, for example, whether they are receiving a medication or a *placebo*.

***élan vital.*** See *life force*.

**empirical.** 1. In philosophy, relating to the view that knowledge is derived from experience rather than from revelation. 2. In science, relating to a conclusion that is derived from observation, rather than from theory.

**endosymbiosis.** The process wherein a smaller cell is engulfed by a larger one and each provides a useful service for the other.

**endothermy.** The condition in which an organism's body temperature is produced and maintained by metabolic heat (see also *exothermy*).

**Enlightenment.** An eighteenth-century philosophical movement that stressed reason. Usually with "the."

**entropy.** See *uncertainty*.

**erosion of genetic diversity.** The loss of genes in a population that has a limited gene pool, in consequence of which individuals with unique genes do not get the opportunity to breed before they die.

**establishment clause.** The clause in the First Amendment to the U.S. Constitution wherein an established religion is prohibited (see also *free-exercise clause*).

**eugenics.** The effort to improve the human race by selective breeding (see also *social Darwinism*).

**eukaryote.** A single- or multi-celled organism whose cell or cells have a nucleus bound by a membrane.

**evangelical.** Usually, a Protestant church that adheres to the doctrine of *biblical inerrancy.*

**event** In probability theory, the outcome of an observation or experiment.

**evolutionary algorithm.** A mathematical *algorithm* or computer program that searches for an optimal solution to a problem by using methods that resemble biological evolution.

**evolutionary arms race.** In biology, an evolutionary struggle involving a series of *adaptations* and counter-adaptations between coevolving genes or organisms.

**evolutionary developmental biology.** The integrated field of evolutionary biology, in which scientists investigate how organisms evolve and change their shape and form, and developmental biology, in which scientists investigate how alterations in gene expression and function lead to changes in body shape and pattern. Also *evo devo*.

**evolutionary synthesis.** The union in the early part of the twentieth century of Mendelian genetics, statistics, and natural selection, among other fields of biology. Also called the *modern synthesis*.

**exothermy.** The condition in which an organism's body temperature is produced and maintained by external sources of heat (see also *endothermy*).

**experimenter bias.** A tendency to interpret experimental results according to one's expectations or preconceptions.

**explanatory filter.** A logical procedure that is said to infer design.

**expression.** In biology, the process by which the specific sequences of DNA bases in a gene are translated into a specific protein product.

**extraneous helium.** In geology, helium that was not generated by radioactive decay within a specimen but rather diffused in from outside the specimen.

**false negative, false positive.** Errors that result from a statistical decision-making procedure. Specifically, a false negative incorrectly fails to show a positive outcome, whereas a false positive incorrectly shows a positive outcome.

**falsifiability.** The principle that any hypothesis must yield conclusions that are sufficiently precise that the hypothesis could be refuted, or falsified, if the conclusions are not supported by observation (see also *testability*).

**falsifiable.** Capable of being disproved, or falsified. Specifically, making falsifiable predictions. Sometimes, testable.

**fine-structure constant.** A dimensionless constant that is related to the strength of the electromagnetic field and is given by a ratio that involves the three fundamental constants, the charge of the electron, Planck's constant, and the speed of light.

**fine-tuning.** The argument that the universe must have been designed because the fundamental constants have apparently been chosen with precise values (see also *anthropic principle*).

**fitness.** In biology, the number of offspring that survive and reproduce.

**flagellum.** A tail-like projection from the membrane of a cell that provides locomotion or adhesion (see also *cilium*).

**flood geology.** The view that the Earth's geological features are the result of the *Noachian* flood.

**fovea.** The region near the center of the *retina* in which the *cone*s are concentrated and that provides the most acute vision. Also fovea centralis.

**fractionation.** In chemistry, the separation into two or more components, for example, by crystallization.

**free living.** The mode of life in which an individual organism is physically independent of any other organism for survival.

**free-exercise clause.** The clause in the First Amendment to the U.S. Constitution wherein freedom of religion is guaranteed (see also *establishment clause*).

**fundamental constant.** One of twenty or more physical constants, such as the mass and charge of the electron, the ratio of the mass of the proton to that of the electron, and the gravitational constant, which cannot be calculated from theory but can only be measured (see also *anthropic principle*, *fine-tuning*).

**gap theory.** The view that the six days of creation represent literal twenty-four-hour days but that a gap between the first and second days accounts for the observed age of the universe (see also *day-age theory*, *old-Earth creationism*).

**gastrulation.** The folding and movement of cells during the development of an embryo that results in the formation of three unique layers of cells called germ layers.

***Gedankenexperiment.*** See *thought experiment*.

**gene.** A segment of DNA that can be indirectly translated into a specific protein with a specific function.

**gene resurrection.** The process by which scientists infer the sequence of ancestral genes and then synthesize them in the lab to generate their protein products.

**genetic drift.** Random change in the frequency of *allele*s in a population that is not the result of *natural selection*.

**genotype.** The entire set of *allele*s for a trait, even those alleles that are not expressed (see also *phenotype*).

**glaucoma.** One of a family of eye diseases that is characterized by damage to the *blind spot* (optic disk) and causes partial or complete blindness.

**God of the gaps.** The view that the existence of God can be deduced from unexplained phenomena, or gaps, in scientific knowledge.

**great synthesis.** See *modern synthesis*.

**group selection.** Reproductive strategies that favor the fitness of members of the *in-group*, often at the expense of the organism employing those strategies (see also *kin selection*).

**heap problem.** In philosophy, the problem of defining when enough particles form a heap. Used to illustrate that some problems do not have clearly demarcated solutions.

**hermeneutics.** The interpretation or methodology of interpretation of religious texts.

**heterotrophic.** Relating to an organism that must consume other organisms or the products of other organisms to obtain nutrients (see also *autotrophic*).

**heuristic.** A formulation that may not be rigorous but provides a guide to solving a specific problem.

**higher criticism.** The use of textual and linguistic analysis to ascertain which parts of the Bible are historically accurate and which parts are not (see also *documentary hypothesis*).

**hill-climbing algorithm.** A mathematical *algorithm* that searches for a solution to an optimization problem by making small, consecutive changes to a trial solution.

**homeobox.** A sequence of approximately 180 DNA bases that codes for a protein region, the *homeodomain*, which recognizes the specific site on a DNA molecule where a developmental control gene begins.

**homeodomain.** A region of about 60 amino acids in a developmental control protein that is responsible for recognizing the region on a developmental control gene where the gene can be turned on (activated).

**homeopathic medicine.** An ineffective medical treatment wherein a disease is treated by administering negligible doses of any substance that in high doses causes symptoms similar to those of the disease.

**homeotic gene.** A gene that controls the location and kind of body segments and parts during development (see also *Hox gene*).

**hominid.** In biology, any member of the family to which humans and great apes belong, including their closely related ancestors.

**homologous.** In biology, pertaining to similar structures among species that share a *common ancestor.*

**horizontal gene flow.** Exchange of genetic material between the three domains of living things (archaea, bacteria, eukarya) early in evolutionary history.

***Hox* gene.** A member of a subgroup of *homeotic genes* that determines which cells in an embryo will become which body part, like legs or antennae.

**hypothesis.** The educated and testable reason for making a specific prediction about natural phenomena.

**inclusion.** In geology, a foreign body trapped in a mineral or a rock.

**infinite regression.** In mathematics, the dependence of successive propositions on other propositions, ad infinitum, so that there is no single starting point.

**information.** The difference between the uncertainties of a meaningful message and a random sequence of the same length.

**in-group.** A group that is united by, for example, common beliefs or common ancestry, and that typically excludes others (see also *out-group*).

**inguinal hernia.** In mammalian males, the protrusion of intestinal material through the inguinal canal, the route created during testicular descent.

**inheritance of acquired characteristics.** A presumed mechanism whereby organisms pass to their descendants certain characteristics that the ancestral organisms acquire during their lifetimes.

**initial conditions.** In physics or mathematics, the presumed value or values of a function or functions at the time that a calculation is to begin.

**intelligent-design creationism.** A modern form of *old-earth creationism;* specifically, the view that evolution must have been guided, at least at times, by a designer, who is presumed to be the Christian God.

**intercessory prayer.** A prayer that asks God (or another entity) to intercede on behalf of another person.

**invagination.** The infolding of cells during the development of an embryo that creates two layers of cells from one layer.

**iris.** The pigmented membrane that is located between the cornea and the lens and regulates the intensity of light incident on the retina. The aperture in the center of the iris is known as the pupil.

**irrational.** Not in accord with reason or logic.

**irreducible complexity.** The argument that certain biological systems are too complex to have evolved from simpler systems (see also *specified complexity*).

**isochron method.** In geology, a method of radiometric dating that does not require knowledge of the initial concentrations of any of the elements.

**isotropic.** Uniform in all directions.

**kin selection.** Reproductive strategies that favor the *fitness* of relatives, often at the expense of the organism employing those strategies (see also *group selection*).

**kind.** In *creation science*, organisms whose existence began at the creation and who have no common ancestors. Corresponds roughly to genus or class.

**knockout.** The process by which a gene is artificially deactivated so that scientists can ascertain the specific function of that gene for an organism, or so that they can replace the gene with another gene.

**laissez-faire.** 1. Non-interference in the affairs of other people. 2. Specifically, a social or economic policy that restricts government regulation of trade and commerce (see also *social Darwinism*).

**lateral transfer.** See *horizontal gene flow*.

**law.** In science, a mathematical or verbal statement, such as the law of gravity, of a well-documented and replicable regularity that is observed under a variety of conditions. A law is generally accepted as a fact, whereas a *theory* is a body of inferences that explains a law (see also *hypothesis*, *prediction*).

**law of conservation of energy.** The principle that energy can be neither created nor destroyed. More properly, the law of conservation of mass-energy.

**law of similars.** See *homeopathic medicine*.

**Lemon test.** The principle that a law involving religion must have a secular purpose, must not primarily advance or inhibit religion, and must not lead to "excessive government entanglement" with religion. All three prongs of the Lemon test must be satisfied if a law is to be constitutional under the establishment clause.

**lens** (of eye). The transparent, biconvex structure that is located behind the *cornea* and helps focus light onto the *retina*.

***lex talionis.*** Law of retaliation, often epitomized as an eye for an eye and a tooth for a tooth.

**life force.** A presumed essence or force that drives inanimate matter to become animate and drives animate matter to evolve.

**logarithmic time.** A nonlinear time scale in which the first day of creation is assumed to have been 8 billion years long, the second day, 4 billion years long, and so on.

**macroevolution.** The sum over time of microevolutionary changes in two separate gene pools, under the influence of large-scale geological and environmental changes, that is at or above the level of speciation (see also *microevolution*).

**magisterium.** The authority to teach any subject.

**magma.** In geology, molten rock beneath the earth' s crust.

**margin of error.** In statistics, the range of probable values within which the true value of a measured quantity is thought to be located. Usually, the true value is thought to lie within the margin of error with 95 percent probability.

**materialism.** The view that all phenomena, including thought, mind, and emotion, can be explained in terms of matter and energy (see also *methodological naturalism, philosophical naturalism*).

**meaningful message.** A nonrandom sequence of bits of *information* that conveys a message, as opposed to a random sequence. Subsumes *complex specified information* and *specified complexity*.

**meta-universe.** See *multiverse*.

**methanogen.** One of a group of microbes in the domain archaea that produce methane gas as a byproduct of metabolism in the absence of oxygen.

**methodological naturalism.** The view that all scientific observations may be explained by natural causes (see also *philosophical naturalism*).

**microevolution.** A change in the frequency of an *allele* in a population over time (see also *macroevolution, speciation*).

**mimicry.** A form of deception in which individuals (the mimics) in a population have evolved a behavior or form that makes them appear indiscernible from another organism (the model) in their environment (see also *camouflage*).

**modern synthesis.** The bringing together in the 1930s of various fields of biology that showed that large-scale evolutionary change could be initiated by *mutation* and *natural selection* (see also *evolutionary synthesis*).

**molecular exploitation.** The production during evolution of new molecules that can combine with ancient molecules, often creating new and complex biological functions and structures.

**multiverse.** A hypothetical set of universes, including our own universe, that together compose everything that exists. Also called meta-universe.

**mutation.** A change in the genetic material of an organism, which sometimes results in a *character* not shared by the parent.

**mythopoetic.** A hypothetical stage of human thinking that predates modern, scientific thinking and is characterized by making myths.

**natural law.** 1. In this book, physical laws that are brought about by natural causes (see also *methodological naturalism*). 2. A body of moral or ethical laws that are thought to derive from nature.

**natural selection.** Differential survival and reproduction, wherein those organisms best adapted to their environment survive and transmit their genetic characters to succeeding generations, whereas those less adapted are eliminated.

**naturalism.** See *methodological naturalism*, *philosophical naturalism*.

**neptunism.** In geology, the theory that the Noachian flood deposited the geological strata, which were then sculpted by the action of the sea (see also *stratification*).

**neutrino.** A subatomic particle, once thought to be massless, that is involved in many nuclear reactions.

**Noachian.** The adjectival form of Noah.

**no-free-lunch theorems.** In computer science, a set of mathematical theorems demonstrating that no single strategy can be applied to all problems or, alternatively, that it is necessary to develop specific strategies for specific problems.

**nonhuman hominid.** People who lived at the time of Adam and Eve and were identical to humans but did not have souls (see, however, *hominid*).

**nonrational.** Not based on reason or logic, but not in discord with reason or logic; intuitive.

**normalization.** In science, dividing one or more data points or data sets by a constant so that they may be compared. Expressing a result as percent change is a way of normalizing a result.

**notochord.** The supportive, rodlike structure found at some point during the development of all members of the phylum Chordata. In vertebrates, this structure is eventually replaced by the backbone.

**Ockham's razor.** The *principle of parsimony*, often expressed as "entities must not be multiplied unnecessarily." Also Occam's razor.

**old-earth creationism.** The view that the universe was created billions of years ago but that the Bible, properly interpreted, describes the creation of the universe (see also *intelligent-design creationism*, *young-earth creationism*).

**ontogeny.** The unique pathway of an organism as it develops from a fertilized egg to an adult.

**ontological naturalism.** See *philosophical naturalism*.

**optic disk.** See *blind spot*.

**optic nerve.** The nerve that originates in the *retina* and carries visual information to the brain.

**organelle.** A structure such as a mitochondrion or a chloroplast that performs a specific function within a cell.

**outbreeding depression.** The condition that the offspring of individuals from two genetically different populations have lower fitness than those of individuals from the same population.

**out-group.** Those excluded by the *in-group*.

**parasite.** An organism that lives in or on another organism, the host, on which it feeds or is sheltered, and contributes nothing to the host but generally does not kill the host (see also *parasitoidism*).

**parasitoidism.** A lethal relationship between a host and a *parasite* wherein the host's body is used as a home or source of nutrition for the offspring of the adult parasite.

**particulate inheritance.** The model proposed by Gregor Mendel that explains the passing of genetic information from parent to offspring as discrete units, now recognized as *genes*, which maintain their unique identity from one generation to the next.

**pathogen.** A biological agent that causes disease in an individual organism.

**phage.** See *bacteriophage*.

**phenotype.** The specific characteristics displayed by an organism that are a result of genes that are expressed (see also *genotype*).

**philosophical naturalism.** The view that there are no causes other than natural causes. This view rules out the supernatural. Distinguished from *methodological naturalism* (see also *materialism*).

**phylogenetic tree.** See *phylogeny*.

**phylogeny.** A branching diagram showing the evolutionary relationships among a group of organisms.

**placebo.** An inactive substance administered as part of a *double-blind test* (see also *experimenter bias*).

**plantigrade.** A form of movement by animals that walk on either two legs or four wherein all the bones of the foot are in contact with the ground during each step (see also *digitigrade, unguligrade*).

**polonium halo.** A visible defect formed in rocks by the radioactive decay of inclusions containing the element polonium.

**polyandry.** The mating system in animals wherein a single female mates with several males.

**polygenic.** Referring to a character or trait that is under the control of many *genes*.

**polywater.** A supposed, now discredited, polymerized form of water which displayed elevated viscosity and boiling point.

**posterior probability.** The probability of occurrence of an event after data about that event are collected.

**precession.** In astronomy, the rotation of the axis of a planet.

**prediction.** The claim that a specific, measurable event, under the influence of specific conditions, will occur in the future (see also *hypothesis*, *theory*).

**premoral behavior.** Cooperative behavior, such as *reciprocal altruism*, among nonhuman animals, especially primates.

**presbyopia.** The inability of the eye to focus sharply on nearby objects owing to a loss of elasticity of the *lens* with increasing age.

**principle of parsimony.** The view that the best solution to a problem is most probably the solution that requires the fewest assumptions or independent variables. Also *Ockham's razor*.

**principle of superposition.** In geology, the principle that layered strata were deposited sequentially (see also *stratification*).

**prior probability.** The probability of occurrence of an event before data about that event are collected.

**prisoner's dilemma.** In game theory, a game in which two isolated prisoners will be freed if neither confesses, but may be forced to minimize their losses and go to jail if they cannot trust each other (see also *ultimatum game*).

**prokaryote.** An organism loosely defined by the absence of a membrane-bound nucleus or *organelles*.

**prong.** See *Lemon test*.

**punctuated equilibrium.** The pattern of evolution seen in the fossil record whereby long periods of stasis (little evolutionary change) are punctuated by the appearance of many new species during a relatively short period.

**quadruped.** A four-legged animal.

**radiometric dating.** A method used to date materials, such as minerals, by examining the relative abundance of radioactive elements and their decay products, and using known decay rates.

**rational.** Based on reason or logic.

**reaction formation.** In psychoanalysis, a defense mechanism in which undesired emotions are replaced by their opposites.

**reciprocal altruism.** A limited form of *altruism* in which one organism provides a benefit to another, in return for a presumed reciprocal benefit to be granted later.

**red shift.** See *Doppler effect*.

**redactor.** 1. An editor. 2. Usually cap. The individual who compiled the five books of Moses (see also *documentary hypothesis*).

**repeated evolution.** The repeated appearance of the same or a similar characteristic among related populations that are genetically isolated from each other, yet are under the influence of same or similar environmental conditions.

**reproductive isolation.** The condition that two populations of the same species can no longer interbreed and thus are prevented from exchanging genes.

**reticulate evolution.** Evolution that is the result of the hybridization of all or part of the genomes of two species.

**retina.** The light-sensitive layer that lines the back of the eyeball and is connected to the brain by means of the *optic nerve* (see also *cone*, *rod*).

**ribosome.** The *organelle* in *prokaryotes* and *eukaryotes* that is the site of protein synthesis.

**rock strata.** See *stratification.*

**rod.** The light-sensitive receptor that is found in the *retina* and provides vision in dim light (see also *cone*).

**sessile.** Permanently attached to a substrate and unable to move about.

**sexual selection.** Preference of either sex to mate with individuals that possess specific external characteristics.

**side information.** Data and facts that may not be used when applying the *explanatory filter.*

**social Darwinism.** The theory that groups or individuals dominate others because of their inherent biological superiority (see also *eugenics*).

**special creation.** 1. The creation of individual species by God. 2. In Roman Catholicism, the creation of human beings; Roman Catholics are required to believe that the human soul was created by God.

**speciation.** The process whereby different populations become reproductively isolated from each other, so that each displays a unique gene pool, and the gene pools are not normally shared (see also *macroevolution, microevolution*); the evolution of one species into two or more species.

**specification.** A feature or a set of features that distinguish a pattern, object, or sequence from related patterns, objects, or sequences; the pattern, object, or sequence for which the *explanatory filter* searches.

**specified complexity.** A *meaningful message* that contains more than 500 bits of information. Also *complex specified information.*

**spectral line.** A single bright line in a spectrum produced by the emission of light at a single wavelength, or a single dark line produced by the absorption of light at a single wavelength.

**spectrum.** In this book, the distribution of intensity emitted by a radiant body and ordered according to its wavelength or frequency.

**spermatogenesis.** The process by which sperm cells are generated by specialized cells in the males of sexually reproducing species.

**spontaneous generation.** The belief that complex organisms developed from inanimate matter. Not to be confused with abiogenesis, or the origin of life.

**steady-state theory.** The now-discredited cosmological theory that the universe, though it is expanding, is static because new matter is continuously created (see also *big bang).*

**strata.** See *stratification.*

**stratification.** In geology, the successive deposition of layers, or strata, of rock, soil, or volcanic ash with characteristics that allow each layer to be distinguished from the others.

**symbiont.** An organism in a relationship with another organism that is usually beneficial, but can also be antagonistic.

***systema naturae.*** The system of nature; the classification system invented by Carolus Linnaeus and published in 1735, wherein all living organisms are organized into succeedingly less inclusive groups, from general kingdoms down to specific genera and species.

**taxonomy.** The classification of organisms into an ordered system according to similarities. Often, the Linnean taxonomy.

**temperature hypothesis.** The hypothesis that the testicles of many mammalian males move into the scrotum outside the body cavity during development because the temperature inside the body cavity is too warm for proper sperm development (see also *inguinal hernia, display hypothesis, testicular descent, training hypothesis*).

**temporal isolation.** A type of reproductive isolation wherein the gene pools of two populations are kept separate because the mating seasons of the two populations occur at different times during the year.

**testability.** The principle that any *hypothesis* must yield conclusions that can be compared against observations, with a view to ascertaining the truth or falsity of that hypothesis (see also *falsifiability*).

**testable.** See *falsifiable.*

**testicular descent.** The process during the development of mammalian males whereby the testicles move toward the pelvic region from inside the body cavity, push through the wall of the abdomen, and into the scrotum, creating the inguinal canal (see also *inguinal hernia, display hypothesis, temperature hypothesis, training hypothesis*).

**tetrapod.** A vertebrate animal characterized by four legs or leglike appendages, and feet.

**theism.** The belief that God (or other deities) is active in the world today (see also *agnosticism, atheism, deism, theistic evolutionism*).

**theistic evolutionism.** The view that evolution is the manner in which God created living organisms; holds that evolution is compatible with classic religious traditions (see also *agnosticism, atheism, deism, theism*).

**theory.** In science, a comprehensive body of knowledge, hypotheses, and deductions that explain a broad range of facts. A term of art not to be confused with a hunch or speculation (see also *law*, *hypothesis*, *prediction*).

**thought experiment.** An experiment that is often not practical but whose outcome can be conceived and sheds light on some scientific problem. Also *Gedankenexperiment*.

***Tiktaalik*.** An extinct vertebrate genus thought to be transitional between fish and tetrapods and characterized, among other features by fish gills, a *tetrapod* lung, a functional wrist joint, a flexible neck, and shoulder and elbow joints.

**tornado in a junkyard.** A specious probabilistic argument designed to show that the evolution of complex organisms is impossible.

**trace fossil.** A fossil of footprints, burrows, and other markings left by organisms typically in moist soil or sand.

**training hypothesis.** The hypothesis that the testicles of many mammalian males move into the scrotum outside the body cavity during development because the scrotum presents the sperm with a hostile environment that, if survived, will result in only the highest quality sperm being available for fertilization (see also *inguinal hernia*, *display hypothesis*, *temperature hypothesis*, *testicular descent*).

**transduction.** The transfer of viral or bacterial genetic material into the genome of a bacterial cell during infection by a *bacteriophage* virus.

**transformation.** The transfer of foreign genetic material into a bacterial cell that causes the cell to express a trait new to that specific bacterium.

**type.** See *kind*.

**ultimatum game.** In economics, a game in which one of two players proposes a division of a sum of money, and the second player either accepts or rejects that proposal (see also *prisoner's dilemma*).

**uncertainty.** A measure of the randomness of a sequence of bits or numbers. Sometimes called *entropy* or *information entropy* by analogy with a related thermodynamic concept.

**unguligrade.** A form of movement specific to hoofed animals whereby they walk on the tips of their toes (see also *digitigrade*, *plantigrade*).

**uniformitarianism.** In geology, the theory that the processes that we see today have been continuing since time immemorial and are in fact the processes that shaped the geological features we observe today (see also *catastrophism*).

**vas deferens.** The tubes in many animals that carry sperm from the site of sperm production to the site of sperm expulsion from the penis.

**vestigial organ.** A rudimentary and nonfunctioning organ that is homologous with a fully developed and functioning organ in preceding generations or species.

**vitreous body.** The clear, gelatinous substance that fills the eyeball between the lens and the retina.

**vitreous humor.** See *vitreous body.*

**von Baer's law.** The principle formulated by Karl Ernst von Baer that embryos of different species share general characteristics early in *development,* but then become separated into more specific groups as they develop more and more specialized structures, and that the embryo of one species never resembles the adult form of another species, only its embryo. Contrast with *biogenic law.*

**Wedge Strategy.** A five-year plan to destroy *materialism* and replace it with an undefined theistic understanding of science.

**young-earth creationism.** The view that the earth was created in approximately its present form at some time in the last 10 to 20 thousand years.

**zircon.** A silicate mineral often used in *radiometric dating.*

# Bibliography

## Books


Ayala, Francisco. 2007. *Darwin's Gift to Science and Religion.* Washington, DC: Joseph Henry Press.

Barbour, Ian G. 1997. *Religion and Science: Historical and Contemporary Issues.* San Francisco: HarperCollins.

Behe, Michael J. 1996. *Darwin's Black Box: The Biochemical Challenge to Evolution.* New York: Touchstone.

———. 2007. *The Edge of Evolution: The Search for the Limits of Darwinism.* New York: Free Press.

Berra, Tim M. 1990. *Evolution and the Myth of Creationism: A Basic Guide to the Facts in the Evolution Debate.* Stanford, CA: Stanford University Press.

Brennan, Edward. 1995. *The Radical Reform of Christianity: a Focus on Catholicism.* Notre Dame, IN: Cross Cultural Publications.

Brockman, John, ed. 2006. *Intelligent Thought: Science versus the Intelligent Design Movement.* New York: Vintage.

Burnie, David. 2003. *Get a Grip on Evolution.* New York: Barnes and Noble.

Carroll, Sean B. 2005. *Endless Forms Most Beautiful: The New Science of Evo Devo.* New York: Norton.

Charlesworth, Brian, and Deborah Charlesworth. 2003. *Evolution: A Very Short Introduction.* Oxford: Oxford University Press.

Collins, Francis. 2006. *The Language of God: A Scientist Presents Evidence for Belief.* New York: Simon & Schuster.

Crews, Frederick. 1996. *The Memory Wars: Freud's Legacy in Dispute.* New York: New York Review of Books.

Dalrymple, G. Brent. 2004. *Ancient Earth, Ancient Skies: The Age of Earth and Its Cosmic Surroundings.* Stanford, CA: Stanford University Press.

Davis, Percival, and Dean H. Kenyon. 1989. *Of Pandas and People: The Central Question of Biological Origins.* Mesquite, TX: Haughton.

Dawkins, Richard. 1987. *The Blind Watchmaker: Why the Evidence of Evolution Reveals a Universe without Design.* New York: Norton.

———. 1996. *Climbing Mount Improbable.* New York: Norton.

De Waal, Frans. 1996. *Good Natured: The Origins of Right and Wrong in Humans and Other Animals.* Cambridge, MA: Harvard University Press.

Dembski, William A. 1999. *Intelligent Design: The Bridge between Science and Theology.* Downers Grove, IL: InterVarsity.

———. 2002. *No Free Lunch: Why Specified Complexity Cannot Be Purchased without Intelligence.* Lanham, MD: Rowman & Littlefield.

Dembski, William A., and Jonathan Wells. 2007. *The Design of Life: Discovering Signs of Intelligence in Biological Systems*. Richardson, TX: Foundation for Thoughts and Ethics.

Dowd, Michael. 2007. *Thank God for Evolution: How the Marriage of Science and Religion Will Transform Your Life and Our World*. New York: Viking.

Feynman, Richard. 1967. *The Character of Physical Law*. Cambridge, MA: MIT Press.

Forrest, Barbara, and Paul R. Gross. 2004. *Creationism's Trojan Horse: The Wedge of Intelligent Design*. Oxford: Oxford University Press.

Friedlander, Michael W. 1995. *At the Fringes of Science*. Boulder, CO: Westview.

Futuyma, Douglas J. 1982. *Science on Trial: The Case for Evolution*. New York: Pantheon.

Godfrey, Steven J., and Christopher R. Smith. 2005. *Paradigms on Pilgrimage: Creationism, Paleontology, and Biblical Interpretation*. Toronto: Clements.

Gonzalez, Guillermo, and Jay Richards. 2004. *The Privileged Planet: How Our Place in the Cosmos Is Designed for Discovery*. Washington, DC: Regnery.

Gould, Stephen Jay. 1977. *Ever Since Darwin: Reflections in Natural History*. New York: Norton.

———. 1999. *Rocks of Ages: Science and Religion in the Fullness of Life*. New York: Ballantine.

Hazen, Robert M. 2005. *Genesis: The Scientific Quest for Life's Origin*. Washington, DC: Joseph Henry Press.

Isaak, Mark. 2007. *The Counter-Creationism Handbook*. Berkeley: University of California Press.

Johnson, Phillip E. 1991. *Darwin on Trial*. Downers Grove, IL: InterVarsity.

Kitcher, Philip. 1982. *Abusing Science: The Case against Creationism*. Cambridge, MA: MIT Press.

———. 2007. *Living with Darwin: Evolution, Design, and the Future of Faith*. Oxford: Oxford University Press.

Kuhn, Thomas S. 1970. *The Structure of Scientific Revolutions*. 2nd ed., enlarged. Chicago: University of Chicago Press.

Kurtz, Paul, ed. 2003. *Science and Religion: Are They Compatible?* Amherst, NY: Prometheus.

Larson, Edward J. 2006. *Evolution: the Remarkable History of a Scientific Theory*. New York: Modern Library.

Lawson, Kristan. 2003. *Darwin and Evolution for Kids: His Life and Ideas, with 21 Activities*. Chicago: Chicago Review Press.

Lebo, Lauri. 2008. *The Devil in Dover: An Insider's Story of Dogma v. Darwin in Small-Town America*. New York: New Press.

Lurquin, Paul F., and Linda Stone. 2007. *Evolution and Religious Creation Myths: How Scientists Respond*. New York: Oxford.

Mayr, Ernst. 2001. *What Evolution Is*. New York: Basic Books.

Meyer, Stephen C., Scott Minnich, Jonathan Moneymaker, Paul A. Nelson, and Ralph Seelke. 2007. *Explore Evolution: The Arguments for and against Neo-Darwinism*. Melbourne and London: Hill House.

Miller, Kenneth R. 1999. *Finding Darwin's God: a Scientist's Search for Common Ground between God and Evolution*. New York: Cliff Street Press.

National Academy of Sciences and Institute of Medicine. 2008. *Science, Evolution, and Creationism*. Washington, DC: National Academies Press.

Numbers, Ronald L. 2006. *The Creationists: From Scientific Creationism to Intelligent Design*. Expanded ed. Cambridge, MA: Harvard University Press.

Perakh, Mark. 2004. *Unintelligent Design*. Amherst, NY: Prometheus.

Peters, Ted, and Martinez Hewlett. 2006. *Can You Believe in God and Evolution? A Guide for the Perplexed*. Nashville, TN: Abingdon.

Petto, Andrew J., and Laurie R. Godfrey, eds. 2008. *Scientists Confront Intelligent Design and Creationism*. 2nd ed. New York: Norton.

Pigluicci, Massimo. 2002. *Denying Evolution: Creationism, Scientism, and the Nature of Science*. Sunderland, MA: Sinauer Associates.

Popper, Karl R. 1965. *Conjectures and Refutations: The Growth of Scientific Knowledge*. New York: Harper and Row.

Ross, Hugh. 1998. *The Genesis Question: Scientific Advances and the Accuracy of Genesis*. Colorado Springs, CO: NavPress.

Ruse, Michael, ed. 1998. *But Is It Science? The Philosophical Question in the Creation/Evolution Controversy*. Amherst, NY: Prometheus.

Ruse, Michael. 2005. *The Evolution-Creation Struggle*. Cambridge, MA: Harvard University Press.

Sarkar, Sahotra. 2007. *Doubting Darwin? Creationist Designs on Evolution*. Malden, MA: Blackwell.

Schilthuizen, Menno. 2001. *Frogs, Flies, and Dandelions: Speciation—The Evolution of New Species*. Oxford: Oxford University Press.

Schroeder, Gerald L. 1990. *Genesis and the Big Bang: The Discovery of Harmony between Modern Science and the Bible*. New York: Bantam.

———. 1997. *The Science of God: The Convergence of Scientific and Biblical Wisdom*. New York: Free Press.

Scott, Eugenie C. 2005. *Evolution versus Creationism: An Introduction*. Berkeley and Los Angeles: University of California Press.

Scott, Eugenie C., and Glenn Branch. 2006. *Not in Our Classrooms: Why Intelligent Design Is Wrong for Our Schools*. Boston: Beacon Press.

Shanks, Niall. 2004. *God, the Devil, and Darwin: A Critique of the Intelligent Design Theory*. Oxford: Oxford University Press.

Shermer, Michael. 2000. *How We Believe: The Search for God in an Age of Science*. New York: Freeman.

———. 2004. *The Science of Good and Evil: Why People Cheat, Gossip, Care, Share, and Follow the Golden Rule*. New York: Henry Holt.

———. 2006. *Why Darwin Matters: The Case against Intelligent Design*. New York: Henry Holt.

Shubin, Neil. 2008. *Your Inner Fish: A Journey into the 3.5-Billion-Year History of the Human Body*. New York: Pantheon.

Skybreak, Ardea. 2006. *The Science of Evolution and the Myth of Creationism*. Chicago: Insight Press.

Smith, Cameron M., and Charles Sullivan. 2007. *The Top 10 Myths about Evolution*. Amherst, NY: Prometheus.

Smolin, Lee. 1997. *The Life of the Cosmos*. Oxford: Oxford University Press.

Stenger, Victor J. 1995. *The Unconscious Quantum: Metaphysics in Modern Physics and Cosmology*. Amherst, NY: Prometheus.

———. 2003. *Has Science Found God? The Latest Results in the Search for Purpose in the Universe*. Amherst, NY: Prometheus.

Strobel, Lee. 2004. *The Case for a Creator: A Journalist Investigates Scientific Evidence That Points toward God*. Grand Rapids, MI: Zondervan.

Swinburne, Richard. 1996. *Is There a God?* Oxford: Oxford University Press.

Wells, Jonathan. 2002. *Icons of Evolution: Science or Myth? Why Much of What We Teach about Evolution Is Wrong*. Washington, DC: Regnery.

Wilson, David Sloan. 2002. *Darwin's Cathedral: Evolution, Religion, and the Nature of Society*. Chicago: University of Chicago Press.

———. 2007. *Evolution for Everyone: How Darwin's Theory Can Change the Way We Think about Our Lives*. New York: Delacorte.

Wright, Robert. 1995. *The Moral Animal: Why We Are the Way We Are: The New Science of Evolutionary Psychology*. New York: Vintage.

Young, Matt. 2001. *No Sense of Obligation: Science and Religion in an Impersonal Universe*. Bloomington, IN: 1stBooks Library.

Young, Matt, and Taner Edis, eds. 2006. *Why Intelligent Design Fails: A Scientific Critique of the New Creationism*. New Brunswick, NJ: Rutgers University Press.

## Blogs and Web Pages

Darwin, Charles. 2002–2008. *The Complete Works of Charles Darwin On Line*. University of Cambridge. http://darwin-online.org.uk/. Date accessed, July 20, 2008.

The National Academies. *Evolution Resources from the National Academies*. http://www.nationalacademies.org/evolution/. Date accessed, July 20, 2008.

National Center for Science Education. "Resources." http://www.natcenscied.org/article.asp. Date accessed, July 20, 2008.

National Evolutionary Synthesis Center, Education and Outreach Group. http://www.nescent.org/eog/. Date accessed, July 20, 2008.

The Panda's Thumb. http://www.pandasthumb.org. Date accessed, July 20, 2008.

RealScience: News and Science Teaching Resources. http://www.realscience.org.uk/. Date accessed, July 21, 2008.

ScienceBlogs—Life Science. http://scienceblogs.com/channel/life-science/. Date accessed, July 20, 2008.

SpringerLink. *Evolution: Education and Outreach*. http://www.springerlink.com/content/120878/. Date accessed, July 20, 2008.

The TalkOrigins Archive: Exploiting the Creation/Evolution Controversy. http://talkorigins.org/. Date accessed, July 20, 2008.

TalkReason. http://www.talkreason.org/index.cfm. Date accessed, July 21, 2008.

University of California at Berkeley. *Understanding Evolution: Your One-Stop Source for Information on Evolution*. http://evolution.berkeley.edu/. Date accessed, July 20, 2008.

Wilkins, John. 2008. "Basic Concepts in Science: A list." http://scienceblogs.com/evolvingthoughts/2008/01/basic_concepts_in_science_a_li.php. *Science Blogs*. Date accessed, July 20, 2008.

Zimmer, Carl. *The Loom*. http://blogs.discovermagazine.com/loom/. Date accessed, July 20, 2008.

Zivkovic, Bora. n.d. "Intro to Life Science." North Carolina Wesleyan College. http://coturnix.wordpress.com/. Date accessed, July 20, 2008. Also posted as Zivkovic, Bora. 2006. "Teaching Biology 101 (to Adults)." *A Blog around the Clock*. http://scienceblogs.com/clock/2006/10/teaching_biology_101_to_adults.php. Date accessed, July 20, 2008.

# Index

## About the Authors

Matt Young is senior lecturer in the Department of Physics at the Colorado School of Mines in Golden, Colorado. He is a coeditor of the book *Why Intelligent Design Fails: A Scientific Critique of the New Creationism*, and the author of books on optics, technical writing, and science and religion. He has published approximately one hundred scientific and technical publications and reports. Until 1999, he was a physicist with the National Institute of Standards and Technology in Boulder, Colorado, and before then held positions at Rensselaer Polytechnic Institute, the Weizmann Institute of Science, and elsewhere. He has been awarded both the gold and silver medals of the Department of Commerce for his work on optical fiber communications, and the Measurement Services Award for developing a standard of fiber diameter. Dr. Young is a fellow emeritus of the Optical Society of America, president of Colorado Citizens for Science, and senior fellow of the Jefferson Center for Science and Religion.

Paul K. Strode teaches biology and a course on research in science for the Boulder (CO) Valley School District, where he is chair of the science review committee. He is also an instructor of ecology and evolutionary biology at the University of Colorado. Before attending graduate school, he taught chemistry and biology in Montana and Washington, where he was awarded a NASA Science Teacher Grant. He has a PhD in ecology and environmental science from the University of Illinois at Urbana–Champaign, an M.Ed. in science education from the University of Washington, and a BS in biology, chemistry, and secondary education from Manchester College in Indiana. He has published a number of technical papers, poster papers, and popular-science papers, and led workshops in several venues, including the National Association of Biology Teachers. While in graduate school, he was awarded the best student poster in the ecology and environment section of the AAAS meeting in 2003 and won the Burgess-Spaeth Scholarship at the University of Illinois. He is now or has been a member of the NABT, the Wildlife Society, the International Society for Ecological Economics, the American Ornithologists Union, and the Ecological Society of America. In 2006, he was interviewed on National Public Radio's "All Things Considered."